AF181774

Computational Nuclear Physics 2

K. Langanke J.A. Maruhn S.E. Koonin
Editors

Computational Nuclear Physics 2

Nuclear Reactions

With 12 Figures and two Program Diskettes

Springer Science+Business Media, LLC

K. Langanke
W.K. Kellogg Radiation Laboratory
California Institute of Technology
Pasadena, CA 91125 USA

J.A. Maruhn
Institut für Theoretische Physik
Universität Frankfurt
Robert-Mayer-Strasse 8-10
6000 Frankfurt 1, Germany

S.E. Koonin
W.K. Kellogg Radiation Laboratory
California Institute of Technology
Pasadena, CA 91125 USA

The series *Computational Physics* was proposed and initiated by K. Langanke.

Library of Congress Cataloging-in-Publication Data
Computational nuclear physics 2 / [edited by] K. Langanke, J.A. Maruhn, S. E. Koonin.
p. cm.
Includes bibliographical references and index.
ISBN 978-1-4613-9337-5
1. Nuclear reactions—Data processing. 2. Hadron interactions—Data processing. 3. Numerical calculations. I. Langanke, K., 1951- . II. Maruhn, J. A. (Joachim A.), 1949- . III. Koonin, Steven E.
QC794.C564 1993
539.7′5′0285—dc20 92-37450

Printed on acid-free paper.

© 1993 Springer Science+Business Media New York
Originally published by Springer-Verlag New York, Inc. in 1993
Softcover reprint of the hardcover 1st edition 1993
Additional material to this book can be downloaded from http://extras.springer.com
All rights reserved. This work may not be translated or copied in whole or in part without the written permission of the publisher (Springer-Verlag New York, Inc., 175 Fifth Avenue, New York, NY 10010, USA), except for brief excerpts in connection with reviews or scholarly analysis. Use in connection with any form of information storage and retrieval, electronic adaptation, computer software, or by similar or dissimilar methodology now known or hereafter developed is forbidden.
The use of general descriptive names, trade names, trademarks, etc., in this publication, even if the former are not especially identified, is not to be taken as a sign that such names, as understood by the Trade Marks and Merchandise Marks Act, may accordingly be used freely by anyone.
The program on the enclosed diskettes is under copyright protection and may not be reproduced without written permission from Springer-Verlag. One copy of the program may be made as a back-up, but all other copies offend copyright law.
Before using the program please consult the technical manuals provided by the manufacturer of the computer—and of any additional plug-in boards—to be used. The authors and publisher accept no legal responsibility for any damage caused by improper use of the instructions and program contained herein. Although the program has been tested with extreme care, we can offer no formal guarantee that it will function correctly.

Production managed by Hal Henglein; manufacturing supervised by Vincent R. Scelta.
Camera-ready copy prepared from the editors' TeX files.

9 8 7 6 5 4 3 2 1

ISBN 978-1-4613-9337-5 ISBN 978-1-4613-9335-1 (eBook)
DOI 10.1007/978-1-4613-9335-1

Preface

Computation is essential to our modern understanding of nuclear systems. Although simple analytical models might guide our intuition, the complexity of the nuclear many-body problem and the ever-increasing precision of experimental results require large-scale numerical studies for a quantitative understanding.

Despite their importance, many nuclear physics computations remain something of a black art. A practicing nuclear physicist might be familiar with one or another type of computation, but there is no way to systematically acquire broad experience. Although computational methods and results are often presented in the literature, it is often difficult to obtain the working codes. More often than not, particular numerical expertise resides in one or a few individuals, who must be contacted informally to generate results; this option becomes unavailable when these individuals leave the field. And while the teaching of modern nuclear physics can benefit enormously from realistic computer simulations, there has been no source for much of the important material.

The present volume, the second of two, is an experiment aimed at addressing some of these problems. We have asked recognized experts in various aspects of computational nuclear physics to codify their expertise in individual chapters. Each chapter takes the form of a brief description of the relevant physics (with appropriate references to the literature), followed by a discussion of the numerical methods used and their embodiment in a FORTRAN code. The chapters also contain sample input and test runs, as well as suggestions for further exploration.

In choosing topics and authors, we have attempted to span the entire field, balancing modern implementations of venerable calculations on the one hand against speculative work at the cutting edge of present research on the other. Thus, in the first volume, concerned primarily with nuclear structure, we included chapters on the shell model and random phase approximation (RPA), as well as on boson and relativistic models. Similarly, this second volume, which deals primarily with nuclear reactions, contains contributions on Hauser–Feshbach and distorted-wave Born approximation (DWBA) calculations, as well as quark models of hadronic interactions. The level of presentation is generally that appropriate for a beginning graduate student who has a working knowledge of both elementary quantum mechanics and computational physics and might have taken (or be taking) a course in nuclear physics.

All of the codes discussed in this book are included on the disk. They have been written to conform to the FORTRAN 77 standard. In some cases, the authors have used subroutines commonly available in libraries (e.g., NAG or IMSL). When this has been done, we have been careful to describe fully the input, output, and purpose of these subroutines, so that the reader can substitute equivalents that might be more readily available. Although the computations are often of a size and length that can be completed on a microcomputer like the IBM PS/2 model 70 in a tolerable amount of time, a workstation or VAX-class machine is generally preferable.

We have edited each chapter only very lightly, preferring instead to exhibit the pleasing diversity of the individual styles. However, we have ensured that each of the chapters is reasonably self-contained and that the codes do run as advertised. There are surely significant opportunities for the synergistic use of the various codes, but we have not explored these in detail.

We suspect that there are at least several uses for the material in this volume. At the simplest level, the codes can be run as black boxes to produce pedagogical illustrations of various nuclear phenomena (for example, the shell-model oscillations in ground-state charge densities or the l-transfer dependence of DWBA cross sections). At a somewhat deeper level, the codes can be used as templates to explore the accuracy and efficiency of various numerical methods within the context of "real" problems. This collection of codes will also undoubtedly find broad use in the research community and might even give our experimental colleagues some appreciation for the theorist's art.

We would like to express our appreciation for the enthusiasm and support that our colleagues have shown for this project, and we offer thanks to Prof. Martin Zirnbauer for his help during the early phases of our work.

Pasadena
March 1992

K. Langanke
J. A. Maruhn
S. E. Koonin

Contents

Contributors

Blüge, Gerhard
Competence Center Informatik
Postfach 1225
W-4470 Meppen, Germany

Faessler, Amand
Institut für Theoretische Physik, Universität Tübingen
W-7400 Tübingen, Germany

Fuchs, Martin
Department of Physics, Oregon State University
Corvallis, Oregon 97331-6507, USA

Gavron, A.
Los Alamos National Laboratory
Los Alamos, New Mexico 87545, USA

Gross, Dieter H.E.
Hahn–Meitner–Institut für Kernforschung
Bereich für Kern- und Strahlenphysik
Postfach 39 01 28, W-1000 Berlin, Germany

Hartnack, Christoph
GSI, Planckstraße 1
D-6100 Darmstadt 11, Germany

Koonin, Steven E.
Kellogg Radiation Laboratory
California Institute of Technology
Pasadena, California 91125, USA

Kruse, Hans G.W.
School of Communication Systems Management
Room 197, RTVC Building
Ohio University
Athens, Ohio 45701, USA

Kunz, P.D.
Nuclear Physics Laboratory
University of Colorado
Boulder, Colorado 80309, USA

Langanke, Karlheinz
Kellogg Radiation Laboratory
California Institute of Technology
Pasadena, California 91125, USA

Lenz, Frieder
Institut für Theoretische Physik, Universität Erlangen-Nürnberg
W-8520 Erlangen, Germany

Machleidt, Ruprecht
Department of Physics
University of Idaho
Moscow, Idaho 83843, USA

Maruhn, Joachim A.
Institut für Theoretische Physik, Johann Wolfgang Goethe Universität
W-6000 Frankfurt, Germany

Müther, Herbert
Institut für Theoretische Physik, Universität Tübingen
W-7400 Tübingen, Germany

Reusch, Hans-Georg
IBM Center für Supercomputing
Tiergartenstraße 15
W-6900 Heidelberg, Germany

Rost, E.
Nuclear Physics Laboratory
University of Colorado
Boulder, Colorado 80309, USA

Sauer, Peter U.
Institut für Theoretische Physik, Universität Hannover
W-3000 Hannover, Germany

Siemens, Philip J.
Department of Physics, Oregon State University
Corvallis, Oregon 97331-6507, USA

Stöcker, Horst
Institut für Theoretische Physik, Johann Wolfgang Goethe Universität
W-6000 Frankfurt, Germany

Stoll, Dieter
Institut für Theoretische Physik
Universität Erlangen-Nürnberg
W-8520 Erlangen, Germany

Straub, U.
Institut für Theoretische Physik, Universität Tübingen
W-7400 Tübingen, Germany

1. One-Boson-Exchange Potentials and Nucleon–Nucleon Scattering

R. Machleidt

1.1 The Meson Theory of Nuclear Forces

The nuclear force has been the heart of nuclear physics ever since the field was born in 1932 with the discovery of the neutron by Chadwick [1.1]. In fact, during the first few decades of nuclear physics, the term nuclear force was often used as synonymous for nuclear physics as a whole. There are fundamental reasons why the nuclear force plays such an outstanding role.

The nucleon–nucleon (NN) interaction has been investigated by a huge army of physicists all over the world for the past 60 years. It is the empirically best-known strong interaction; in fact, for no other strong interaction has a comparable amount of experimental data been accumulated. Attempts to understand the nature of this force have spawned the sister discipline of particle physics.

The interaction between two nucleons is the basis for all of nuclear physics. Indeed, the traditional goal of nuclear physics is to understand the properties of atomic nuclei in terms of the "basic" interaction between pairs of nucleons. With the onset of quantumchromodynamics (QCD), it became clear that this interaction is not fundamental. However, even today as a first approach towards any nuclear-structure problem, one assumes the nucleons to be elementary particles. The failure or success of this approach may then teach us something about the relevance of subnuclear degrees of freedom for the nuclear-structure problem under consideration.

Historically, it turned out to be a terrible task to describe the nuclear force just phenomenologically. Nowadays, an acceptable nuclear potential should be derived from theoretical concepts. Ideally, these concepts should be based on QCD. Unfortunately, on the QCD scale, nuclear physics deals with low-energy processes for which QCD is highly non-perturbative. This implies horrific mathematical problems that are far from being solved.

The oldest attempt to explain the nature of the nuclear force is due to Yukawa [1.2]. According to his theory, massive bosons (mesons) mediate the interaction between two nucleons. Although, in the light of QCD, this is not a theory in the true sense of the word, the meson exchange concept is even today the only one capable of yielding a quantitative nucleon–nucleon potential. This fact is not as puzzling as it may appear at first glance: at low energies, the confinement of quarks into (colorless) hadrons, mesons, and baryons is one of the most fundamental predictions of QCD. Of course, the couplings between hadrons are not fundamental. However, they cannot be arbitrary, since effective hadron–hadron couplings also have to satisfy the symmetries of strong interactions.

There are essentially two ways to derive the nuclear force from meson theory: dispersion relations and field theory. The Stony Brook [1.3] and Paris [1.4] groups pursued the former concept. Dispersion relations determine the on-shell NN T-matrix from empirical information on π–N scattering. Since an on-shell T-matrix is not a potential, certain *ad hoc* prescriptions have to be introduced to define the potential. Furthermore, dispersion relations present formidable mathematical complexity. Therefore, the Paris potential [1.5] was eventually parametrized in terms of simple Yukawa functions of multiples of the pion mass. To fit the NN data correctly, a very large phenomenological short-range potential also had to be introduced. Our approach, on the other hand, is field-theoretic. This has the advantage that the potential is *a priori* well-defined on- and off-the-energy-shell.

A truly consistent field-theoretic meson model for the nuclear force is the Bonn full model [1.6]. Besides the one-meson-exchange contributions, it contains a comprehensive model for the two-pion exchange including virtual isobar excitation. Moreover, the important irreducible diagrams of π *and* ρ meson exchange are also taken into account. Finally, classes of 3π and 4π exchanges are considered. Among the multi-meson-exchange contributions, large cancellations occur suggesting some kind of convergence of the diagrammatic expansion. An even more important observation from the Bonn full model is that the sum of all two-pion and higher-order diagrams can, to a very good approximation, be replaced by the exchange of a scalar-isoscalar boson with a mass of about 550 MeV.

The development of the Bonn full model was necessary to assess systematically the range of validity of the meson-exchange concept for the nuclear force. Thus, the Bonn model provides a benchmark against which QCD-inspired work on the NN interaction can be tested. Furthermore, this model provides a basis for a comprehensive discussion of relativistic effects, medium effects, and many-body forces to be expected in the nuclear many-body problem. The medium effects on the nuclear force when inserted into nuclear matter have been calculated thoroughly. They have turned out to be repulsive, the largest effect coming from intermediate isobars. On the other hand, isobars also give rise to many-body forces that are attractive. Thus, there are large cancellations between these two classes of many-body forces/effects, and the net contribution is very small [1.7]. Relativistic effects, however, may play a non-negligible role in the nuclear many-body problem [1.7, 1.8].

In view of these results, the relativistic one-boson-exchange (OBE) model for the NN force is sufficient for many problems of nuclear-structure physics. This model takes into account only single exchanges of bosons with masses below 1 GeV. The model, which we will present below in more detail, consists of six bosons of which the following four play the major role:

- The pseudo-scalar pion with a mass of about 138 MeV. It is the lightest meson and provides the long-range part of the potential and most of the tensor force.
- The ρ meson, a 2π P-wave resonance of about 770 MeV. Its major effect is to cut down the tensor force provided by the pion at short range.
- The ω meson, a 3π resonance of 783 MeV and spin 1. It creates a strong repulsive central force of short range ("repulsive core") and the nuclear spin–orbit force.
- The (fictitious) σ boson of about 550 MeV. It provides the intermediate-range attraction necessary for nuclear binding and can be understood as a simulation of the correlated S-wave 2π-exchange.

The features of the various one-boson exchanges listed can be derived analytically in the non-relativistic approximation to the Feynman amplitudes for these processes [1.7]. However, these non-relativistic simplifications should be seen basically as a pedagogical matter: they yield a local potential that is easy to visualize. It has turned out that for a quantitative description of the NN scattering data the subtleties of the full-field-theoretic Feynman amplitude are important. In other words, non-localities play a subtle, but important, role. The local, non-relativistic expressions do not allow for a very quantitative description of the NN data unless additional phenomenological terms are introduced. This, however, increases substantially the number of parameters and thus destroys the beauty and self-consistency of the model.

The relativistic OBE model is able to describe the NN data with the parameters of just six mesons. The χ^2 for the fit of all NN data below 300 MeV laboratory energy is as good as the one of semiphenomenological models that use in the order of 200 parameters, such as the Paris potential [1.5].

The relativistic OBE potentials are suitable for application in relativistic as well as non-relativistic nuclear-structure physics (see the discussion of the appropriate equations below).

In the following sections, we will be mainly concerned with the mathematical and numerical aspects of NN scattering and one-boson-exchange forces. A broad introduction to the physics of the field is given in [1.7]. Ultimately, we present a computer code for one-boson-exchange potentials and one for the computation of phase shifts of NN scattering. All calculations are performed in momentum space. The advantage of working in momentum space is that local and non-local potentials can be handled mathematically in the same way. Thus, the calculations with non-local potentials are as easy as those with local ones. Moreover, the scattering equation in momentum space is an integral equation, which is easy to solve numerically. In coordinate space, the scattering equation is a differential equation, the numerical analysis of which is always more involved.

1.2 Relativistic Two-Nucleon Scattering

1.2.1 Covariant Equations

Two-nucleon scattering is described covariantly by the Bethe–Salpeter (BS) equation [1.9]. In operator notation, it may be written as

$$\mathcal{T} = \mathcal{V} + \mathcal{V}\mathcal{G}\mathcal{T}, \tag{1.1}$$

with $\mathcal{T}$ the invariant amplitude for the two-nucleon scattering process, $\mathcal{V}$ the sum of all connected two-particle irreducible diagrams, and $\mathcal{G}$ the relativistic two-nucleon propagator. As this four-dimensional integral equation is very difficult to solve, so-called three-dimensional reductions have been proposed, which are more amenable to numerical solution. Furthermore, it has been shown by Gross [1.10] that the full BS equation in ladder approximation (that is, the kernel $\mathcal{V}$ is restricted to the exchange of single particles, as, for example, in the OBE model) does not have the correct one-body limit (that is, when one of the particles becomes very massive), while a large family of three-dimensional quasi-potential equations does. These approximations to the BS equation are also covariant and satisfy relativistic elastic unitarity. However, the three-dimensional reduction is not unique, and in principle, infinitely many choices exist. Typically they are derived by replacing (1.1) with two coupled equations [1.11]:

$$\mathcal{T} = \mathcal{W} + \mathcal{W}g\mathcal{T} \tag{1.2}$$

and

$$\mathcal{W} = \mathcal{V} + \mathcal{V}(\mathcal{G} - g)\mathcal{W}, \tag{1.3}$$

where g is a covariant three-dimensional propagator with the same elastic unitarity cut as $\mathcal{G}$ in the physical region. In general, the second term on the right-hand side of (1.3) is dropped to obtain a true simplification of the problem.

More explicitly, the BS equation for an arbitrary frame reads (notation and conventions as in [1.12])

$$\mathcal{T}(q';q|P) = \mathcal{V}(q';q|P) + \int \mathrm{d}^4k\, \mathcal{V}(q';k|P)\, \mathcal{G}(k|P)\, \mathcal{T}(k;q|P), \tag{1.4}$$

with

$$\mathcal{G}(k|P) = \frac{\mathrm{i}}{2\pi} \frac{1}{(\frac{1}{2}\not{P} + \not{k} - M + \mathrm{i}\epsilon)^{(1)}} \frac{1}{(\frac{1}{2}\not{P} - \not{k} - M + \mathrm{i}\epsilon)^{(2)}} \tag{1.5}$$

$$= \frac{\mathrm{i}}{2\pi} \left[\frac{\frac{1}{2}\not{P} + \not{k} + M}{(\frac{1}{2}P + k)^2 - M^2 + \mathrm{i}\epsilon}\right]^{(1)} \left[\frac{\frac{1}{2}\not{P} - \not{k} + M}{(\frac{1}{2}P - k)^2 - M^2 + \mathrm{i}\epsilon}\right]^{(2)}, \tag{1.6}$$

where q, k, and q' are the initial, intermediate, and final relative four-momenta, respectively, and $P = (P_0, \boldsymbol{P})$ is the total four-momentum. For example, in the initial state we have $q = \frac{1}{2}(p_1 - p_2), P = p_1 + p_2$, and $p_{1/2} = \frac{1}{2}P \pm q$, with p_1 and p_2 the individual four-momenta of particles 1 and 2. In the center-of-mass (c.m.) frame, we will have $P = (\sqrt{s}, \mathbf{0})$, with $\sqrt{s}$ the total energy. For all four-momenta, our notation is $k = (k_0, \boldsymbol{k})$, $\not{k} \equiv \gamma^\mu k_\mu$. M denotes the nucleon mass. The superscripts in (1.6) refer to particles (1) and (2). At this stage, $\mathcal{M}, \mathcal{V}$, and $\mathcal{G}$ are operators in spinor space, that is, they are 16×16 matrices that, when sandwiched between Dirac spinors, yield the corresponding matrix elements.

It is common to the derivation of all three-dimensional reductions that the time component of the relative momentum is fixed in some covariant way, so that it no longer appears as an independent variable in the propagator.

Following Blankenbecler and Sugar (BbS) [1.13], one possible choice for g is (stated in manifestly covariant form for an arbitrary frame)

$$\begin{aligned} g_{\rm BbS}(k,s) = &-\int_{4M^2}^{\infty} \frac{\mathrm{d}s'}{s'-s-\mathrm{i}\epsilon}\, \delta^{(+)}[(\tfrac{1}{2}P'+k)^2 - M^2] \\ &\times \delta^{(+)}[(\tfrac{1}{2}P'-k)^2 - M^2] \\ &\times (\tfrac{1}{2}\not{P}' + \not{k} + M)^{(1)} (\tfrac{1}{2}\not{P}' - \not{k} + M)^{(2)}, \end{aligned} \tag{1.7}$$

with $\delta^{(+)}$ indicating that only the positive energy root of the argument of the δ-function is to be included; $P^2 = s$ and $P' \equiv \frac{\sqrt{s'}}{\sqrt{s}}P$. By construction, the propagator $g_{\rm BbS}$ has the same imaginary part as $\mathcal{G}$ and therefore preserves the unitarity relation satisfied by $\mathcal{T}$. In the c.m. frame, integration yields

$$g_{\rm BbS}(k,s) = \delta(k_0)\, \bar{g}_{\rm BbS}(\boldsymbol{k}, s) \tag{1.8}$$

with

$$\bar{g}_{\rm BbS}(\boldsymbol{k}, s) = \frac{M^2}{E_k} \frac{\Lambda_+^{(1)}(\boldsymbol{k})\, \Lambda_+^{(2)}(-\boldsymbol{k})}{\frac{1}{4}s - E_k^2 + \mathrm{i}\epsilon}, \tag{1.9}$$

where

$$\Lambda_+^{(i)}(\boldsymbol{k}) = \left(\frac{\gamma^0 E_k - \boldsymbol{\gamma}\cdot\boldsymbol{k} + M}{2M}\right)^{(i)} \tag{1.10}$$

$$= \sum_{\lambda_i} |u(\boldsymbol{k}, \lambda_i)\rangle\langle\bar{u}(\boldsymbol{k}, \lambda_i)| \tag{1.11}$$

represents the positive-energy projection operator for nucleon i with $u(\boldsymbol{k})$ a positive-energy Dirac spinor of momentum $\boldsymbol{k}$, $\bar{u} \equiv u^\dagger\gamma^0$. λ_i denotes either the helicity or the spin projection of the respective nucleon, and $E_k = \sqrt{M^2 + \boldsymbol{k}^2}$. The projection operators imply that virtual anti-nucleon contributions are suppressed. These contributions have been shown to be small if the pseudovector coupling is used for the pion [1.14].

Using the approximation $\mathcal{W} = \mathcal{V}$, we obtain the explicit form of (1.2) replacing $\mathcal{G}$ by $g_{\rm BbS}$ in (1.4). This yields, in the c.m. frame,

$$\begin{aligned} &\mathcal{T}(0,\boldsymbol{q}';0,\boldsymbol{q}|\sqrt{s}) \\ &= \mathcal{V}(0,\boldsymbol{q}';0,\boldsymbol{q}) + \int \mathrm{d}^3k\, \mathcal{V}(0,\boldsymbol{q}';0,\boldsymbol{k})\, \bar{g}_{\rm BbS}(\boldsymbol{k},s)\, \mathcal{T}(0,\boldsymbol{k};0,\boldsymbol{q}|\sqrt{s}). \end{aligned} \tag{1.12}$$

Note that four-momentum is conserved at each vertex, and that in the initial state the nucleons are on their mass-shell, therefore $q = (0, \boldsymbol{q})$. The total c.m. energy is

$$\sqrt{s} = 2E_q = 2\sqrt{M^2 + \boldsymbol{q}^2}. \tag{1.13}$$

With this we obtain, simplifying our notation,

$$\mathcal{T}(\boldsymbol{q}', \boldsymbol{q}) = \mathcal{V}(\boldsymbol{q}', \boldsymbol{q}) + \int \mathrm{d}^3k\, \mathcal{V}(\boldsymbol{q}', \boldsymbol{k})\, \frac{M^2}{E_k}\, \frac{\Lambda_+^{(1)}(\boldsymbol{k})\, \Lambda_+^{(2)}(-\boldsymbol{k})}{\boldsymbol{q}^2 - \boldsymbol{k}^2 + \mathrm{i}\epsilon}\, \mathcal{T}(\boldsymbol{k}, \boldsymbol{q}). \tag{1.14}$$

Taking matrix elements between positive-energy spinors (see Sect. 1.3.1 where this is done explicitly for $\mathcal{V}$) yields an equation for the scattering amplitude

$$T(\boldsymbol{q}', \boldsymbol{q}) = V(\boldsymbol{q}', \boldsymbol{q}) + \int \mathrm{d}^3k\, V(\boldsymbol{q}', \boldsymbol{k})\, \frac{M^2}{E_k}\, \frac{1}{\boldsymbol{q}^2 - \boldsymbol{k}^2 + \mathrm{i}\epsilon}\, T(\boldsymbol{k}, \boldsymbol{q}), \tag{1.15}$$

where, for the moment, spin (or helicity) and isospin indices are suppressed (these details will be discussed in Sect. 1.2.2).

Defining

$$\hat{T}(\boldsymbol{q}', \boldsymbol{q}) = \sqrt{\frac{M}{E_{q'}}}\, T(\boldsymbol{q}', \boldsymbol{q})\, \sqrt{\frac{M}{E_q}} \tag{1.16}$$

and

$$\hat{V}(\boldsymbol{q}', \boldsymbol{q}) = \sqrt{\frac{M}{E_{q'}}}\, V(\boldsymbol{q}', \boldsymbol{q})\, \sqrt{\frac{M}{E_q}}, \tag{1.17}$$

which have become known as "minimal relativity" [1.15], we can rewrite (1.15) as

$$\hat{T}(\boldsymbol{q}', \boldsymbol{q}) = \hat{V}(\boldsymbol{q}', \boldsymbol{q}) + \int \mathrm{d}^3k\, \hat{V}(\boldsymbol{q}', \boldsymbol{k})\, \frac{M}{\boldsymbol{q}^2 - \boldsymbol{k}^2 + \mathrm{i}\epsilon}\, \hat{T}(\boldsymbol{k}, \boldsymbol{q}), \tag{1.18}$$

which has the form of the non-relativistic Lippmann–Schwinger equation. A potential defined within an equation that is formally identical to the (non-relativistic) Lippmann–Schwinger equation can then be applied in conventional (non-relativistic) nuclear-structure physics. This is the practical relevance of (1.16)–(1.18). On the other hand, for relativistic nuclear-structure physics, (1.15) is the starting point.

The BbS propagator is the most widely used approximation. Another choice that has been frequently applied is the version suggested by Thompson [1.16], which reads, in manifestly covariant form for an arbitrary frame,

$$\begin{aligned} g_{\mathrm{Th}}(k, s) = &-\int_{2M}^{\infty} \frac{\mathrm{d}\sqrt{s'}}{\sqrt{s'} - \sqrt{s} - \mathrm{i}\epsilon}\, \delta^{(+)}[(\tfrac{1}{2}P' + k)^2 - M^2] \\ &\times \delta^{(+)}[(\tfrac{1}{2}P' - k)^2 - M^2] \\ &\times (\tfrac{1}{2}\, \not{P}' + \not{k} + M)^{(1)} (\tfrac{1}{2}\, \not{P}' - \not{k} + M)^{(2)}. \end{aligned} \tag{1.19}$$

In the c.m. frame, integration yields

$$g_{\mathrm{Th}}(k,s)=\delta(k_0)\,\frac{M^2}{2E_k^2}\,\frac{\Lambda_+^{(1)}(\boldsymbol{k})\,\Lambda_+^{(2)}(-\boldsymbol{k})}{\frac{1}{2}\sqrt{s}-E_k+\mathrm{i}\epsilon}\,. \tag{1.20}$$

The corresponding T-matrix equation is

$$T(\boldsymbol{q}',\boldsymbol{q})=V(\boldsymbol{q}',\boldsymbol{q})+\int \mathrm{d}^3k\,V(\boldsymbol{q}',\boldsymbol{k})\,\frac{M^2}{2E_k^2}\,\frac{1}{E_q-E_k+\mathrm{i}\epsilon}\,T(\boldsymbol{k},\boldsymbol{q}). \tag{1.21}$$

Again, we may introduce some special definitions. In the case of the Thompson equation, it is convenient to define

$$\check{T}(\boldsymbol{q}',\boldsymbol{q})=\frac{M}{E_{q'}}\,T(\boldsymbol{q}',\boldsymbol{q})\,\frac{M}{E_q} \tag{1.22}$$

and

$$\check{V}(\boldsymbol{q}',\boldsymbol{q})=\frac{M}{E_{q'}}\,V(\boldsymbol{q}',\boldsymbol{q})\,\frac{M}{E_q}\,. \tag{1.23}$$

With this, we can rewrite (1.21) as

$$\check{T}(\boldsymbol{q}',\boldsymbol{q})=\check{V}(\boldsymbol{q}',\boldsymbol{q})+\int \mathrm{d}^3k\,\check{V}(\boldsymbol{q}',\boldsymbol{k})\frac{1}{2E_q\quad 2E_k+\mathrm{i}\epsilon}\check{T}(\boldsymbol{k},\boldsymbol{q}), \tag{1.24}$$

which looks like a Lippmann–Schwinger with relativistic energies in the propagator.

Notice that the quantity T in (1.15) and (1.21) is invariant, while $\hat{T}$ of (1.18) and $\check{T}$ of (1.22) are not. $\hat{T}$ is equivalent to the familiar non relativistic T-matrix.

The relationship of the invariant T to the S-matrix is

$$\langle p_1'p_2'|S|p_1p_2\rangle=\langle p_1'p_2'|p_1p_2\rangle-2\pi\mathrm{i}\delta^{(4)}(p_1'+p_2'-p_1-p_2)\frac{M^2}{E_q^2}T(\boldsymbol{q}',\boldsymbol{q}) \tag{1.25}$$

with p_i the initial and p_i' $(i=1,2)$ the final four-momenta of the two interacting nucleons. The normalization is $\langle p'|p\rangle=\delta^{(3)}(\boldsymbol{p}'-\boldsymbol{p})$.

1.2.2 The R-Matrix Equation and Helicity State Basis

On a computer, real analysis is much faster than complex analysis. It is therefore desirable to deal with real quantities whenever possible. The scattering amplitude below particle production threshold can be expressed in terms of the real R-matrix (better known as the "K-matrix") which is defined by

$$T=R-\mathrm{i}\pi R\delta(E-H_0)T. \tag{1.26}$$

The equation for the real $\hat{R}$-matrix corresponding to the complex $\hat{T}$-matrix of (1.18) is

$$\hat{R}(\boldsymbol{q}',\boldsymbol{q}) = \hat{V}(\boldsymbol{q}',\boldsymbol{q}) + \mathcal{P}\int d^3k\, \hat{V}(\boldsymbol{q}',\boldsymbol{k})\, \frac{M}{\boldsymbol{q}^2-\boldsymbol{k}^2}\, \hat{R}(\boldsymbol{k},\boldsymbol{q}), \tag{1.27}$$

where $\mathcal{P}$ denotes the principal value.

Now, we also need to consider the spin of the nucleons explicitly. The easiest way to treat the spin projections of spin-$\frac{1}{2}$ particles in a covariant way is to use the helicity representation. In our further developments, we will therefore use a helicity state basis. Our presentation will be relatively brief, emphasizing the formulas used in the computer codes. A more detailed derivation is given in Appendix C of [1.6], which is based upon [1.17,18].

The helicity λ_i of particle i (with $i = 1$ or 2) is the eigenvalue of the helicity operator $\frac{1}{2}\boldsymbol{\sigma}_i \cdot \boldsymbol{p}_i/|\boldsymbol{p}_i|$, that is, $\pm\frac{1}{2}$.

Using helicity states, the $\hat{R}$-matrix equation reads, after partial wave decomposition,

$$\begin{aligned}\langle\lambda_1'\lambda_2'|\hat{R}^J(q',q)|\lambda_1\lambda_2\rangle = {} & \langle\lambda_1'\lambda_2'|\hat{V}^J(q',q)|\lambda_1\lambda_2\rangle \\ & + \sum_{h_1,h_2} \mathcal{P}\int_0^\infty dk\, k^2 \frac{M}{q^2-k^2} \langle\lambda_1'\lambda_2'|\hat{V}^J(q',k)|h_1h_2\rangle \\ & \times \langle h_1h_2|\hat{R}^J(k,q)|\lambda_1\lambda_2\rangle,\end{aligned} \tag{1.28}$$

where J denotes the total angular momentum of the two nucleons. Here and throughout the rest of this chapter, momenta denoted by non-bold letters are absolute three-momenta, so that $q \equiv |\boldsymbol{q}|$, $k \equiv |\boldsymbol{k}|$, etc.; h_1 and h_2 are the helicities in intermediate states for nucleons 1 and 2, respectively. Equation (1.28) is a system of coupled integral equations that needs to be solved to obtain the desired matrix elements of $\hat{R}^J$.

Ignoring antiparticles, there are $4 \times 4 = 16$ helicity amplitudes for $\hat{R}^J$. However, time-reversal invariance, parity conservation, and the fact that we are dealing with two identical fermions imply that only six amplitudes are independent. For these six amplitudes, we choose the following set:

$$\begin{aligned}\hat{R}_1^J(q',q) &\equiv \langle ++|\hat{R}^J(q',q)|++\rangle, \\ \hat{R}_2^J(q',q) &\equiv \langle ++|\hat{R}^J(q',q)|--\rangle, \\ \hat{R}_3^J(q',q) &\equiv \langle +-|\hat{R}^J(q',q)|+-\rangle, \\ \hat{R}_4^J(q',q) &\equiv \langle +-|\hat{R}^J(q',q)|-+\rangle, \\ \hat{R}_5^J(q',q) &\equiv \langle ++|\hat{R}^J(q',q)|+-\rangle, \\ \hat{R}_6^J(q',q) &\equiv \langle +-|\hat{R}^J(q',q)|++\rangle,\end{aligned} \tag{1.29}$$

where $\pm$ stands for $\pm\frac{1}{2}$. Notice that

$$\hat{R}_5^J(q',q) = \hat{R}_6^J(q,q'). \tag{1.30}$$

We now have six coupled equations. To partially decouple this system, it is useful to introduce the following linear combinations of helicity amplitudes:

$${}^0\hat{R}^J \equiv \hat{R}_1^J - \hat{R}_2^J,$$

$$
\begin{aligned}
{}^{1}\hat{R}^{J} &\equiv \hat{R}^{J}_{3} - \hat{R}^{J}_{4},\\
{}^{12}\hat{R}^{J} &\equiv \hat{R}^{J}_{1} + \hat{R}^{J}_{2},\\
{}^{34}\hat{R}^{J} &\equiv \hat{R}^{J}_{3} + \hat{R}^{J}_{4},\\
{}^{55}\hat{R}^{J} &\equiv 2\hat{R}^{J}_{5},\\
{}^{66}\hat{R}^{J} &\equiv 2\hat{R}^{J}_{6}.
\end{aligned} \tag{1.31}
$$

We also introduce corresponding definitions for $\hat{V}^J$. (Of course, analogous definitions exist for $R, \check{R}, V,$ and $\check{V}$.) Using these definitions, (1.28) decouples into the following three subsystems of integral equations:
Spin singlet

$$
{}^{0}\hat{R}^{J}(q',q) = {}^{0}\hat{V}^{J}(q',q) + \mathcal{P}\int_0^\infty \mathrm{d}k\, k^2 \frac{M}{q^2-k^2}\, {}^{0}\hat{V}^{J}(q',k)\, {}^{0}\hat{R}^{J}(k,q); \tag{1.32}
$$

Uncoupled spin triplet

$$
{}^{1}\hat{R}^{J}(q',q) = {}^{1}\hat{V}^{J}(q',q) + \mathcal{P}\int_0^\infty \mathrm{d}k\, k^2 \frac{M}{q^2-k^2}\, {}^{1}\hat{V}^{J}(q',k)\, {}^{1}\hat{R}^{J}(k,q); \tag{1.33}
$$

Coupled triplet states

$$
\begin{aligned}
{}^{12}\hat{R}^{J}(q',q) = {}^{12}\hat{V}^{J}(q',q) + \mathcal{P}\int_0^\infty \mathrm{d}k\, k^2 \frac{M}{q^2-k^2}[\,&{}^{12}\hat{V}^{J}(q',k)\, {}^{12}\hat{R}^{J}(k,q)\\
&+ {}^{55}\hat{V}^{J}(q',k)\, {}^{66}\hat{R}^{J}(k,q)],\\
{}^{34}\hat{R}^{J}(q',q) = {}^{34}\hat{V}^{J}(q',q) + \mathcal{P}\int_0^\infty \mathrm{d}k\, k^2 \frac{M}{q^2-k^2}[\,&{}^{34}\hat{V}^{J}(q',k)\, {}^{34}\hat{R}^{J}(k,q)\\
&+ {}^{66}\hat{V}^{J}(q',k)\, {}^{55}\hat{R}^{J}(k,q)],\\
{}^{55}\hat{R}^{J}(q',q) = {}^{55}\hat{V}^{J}(q',q) + \mathcal{P}\int_0^\infty \mathrm{d}k\, k^2 \frac{M}{q^2-k^2}[\,&{}^{12}\hat{V}^{J}(q',k)\, {}^{55}\hat{R}^{J}(k,q)\\
&+ {}^{55}\hat{V}^{J}(q',k)\, {}^{34}\hat{R}^{J}(k,q)],\\
{}^{66}\hat{R}^{J}(q',q) = {}^{66}\hat{V}^{J}(q',q) + \mathcal{P}\int_0^\infty \mathrm{d}k\, k^2 \frac{M}{q^2-k^2}[\,&{}^{34}\hat{V}^{J}(q',k)\, {}^{66}\hat{R}^{J}(k,q)\\
&+ {}^{66}\hat{V}^{J}(q',k)\, {}^{12}\hat{R}^{J}(k,q)].
\end{aligned} \tag{1.34}
$$

More common in nuclear physics is the representation of two-nucleon states in terms of an $|LSJM\rangle$ basis, where S denotes the total spin, L the total orbital angular momentum, and J the total angular momentum with projection M. In this basis, we will denote the $\hat{R}$ matrix elements by $\hat{R}^{JS}_{L',L} \equiv \langle L'SJM|\hat{R}|LSJM\rangle$. These are obtained from the helicity state matrix elements by the following unitary transformations:
Spin singlet

$$
\hat{R}^{J0}_{J,J} = {}^{0}\hat{R}^{J}; \tag{1.35}
$$

Uncoupled spin triplet

$$\hat{R}^{J1}_{J,J} = {}^1\hat{R}^J; \tag{1.36}$$

Coupled triplet states

$$\begin{aligned}
\hat{R}^{J1}_{J-1,J-1} &= \frac{1}{2J+1} \\
&\times \left[J\,{}^{12}\hat{R}^J + (J+1)\,{}^{34}\hat{R}^J + \sqrt{J(J+1)}({}^{55}\hat{R}^J + {}^{66}\hat{R}^J) \right], \\
\hat{R}^{J1}_{J+1,J+1} &= \frac{1}{2J+1} \\
&\times \left[(J+1)\,{}^{12}\hat{R}^J + J\,{}^{34}\hat{R}^J - \sqrt{J(J+1)}({}^{55}\hat{R}^J + {}^{66}\hat{R}^J) \right], \\
\hat{R}^{J1}_{J-1,J+1} &= \frac{1}{2J+1} \\
&\times \left[\sqrt{J(J+1)}({}^{12}\hat{R}^J - {}^{34}\hat{R}^J) - J\,{}^{55}\hat{R}^J + (J+1)\,{}^{66}\hat{R}^J \right], \\
\hat{R}^{J1}_{J+1,J-1} &= \frac{1}{2J+1} \\
&\times \left[\sqrt{J(J+1)}({}^{12}\hat{R}^J - {}^{34}\hat{R}^J) + (J+1)\,{}^{55}\hat{R}^J - J\,{}^{66}\hat{R}^J \right].
\end{aligned} \tag{1.37}$$

Analogous transformations exist for $\hat{V}, R, \check{R}, V,$ and $\check{V}$. Notice that the transformation matrix is symmetric, which implies that the inverse transformation has the same matrix.

Instead of solving the coupled system in the form of (1.34), one can also first apply the transformation (1.37) to $\hat{V}$ and $\hat{R}$ in (1.34), yielding a system of four coupled integral equations for $\hat{R}$ in LSJ representation:

$$\begin{aligned}
\hat{R}^{J1}_{++}(q',q) &= \hat{V}^{J1}_{++}(q',q) + \mathcal{P}\int_0^\infty \mathrm{d}k\, k^2 \frac{M}{q^2-k^2}[\,\hat{V}^{J1}_{++}(q',k)\,\hat{R}^{J1}_{++}(k,q) \\
&\quad + \hat{V}^{J1}_{+-}(q',k)\,\hat{R}^{J1}_{-+}(k,q)], \\
\hat{R}^{J1}_{--}(q',q) &= \hat{V}^{J1}_{--}(q',q) + \mathcal{P}\int_0^\infty \mathrm{d}k\, k^2 \frac{M}{q^2-k^2}[\,\hat{V}^{J1}_{--}(q',k)\,\hat{R}^{J1}_{--}(k,q) \\
&\quad + \hat{V}^{J1}_{-+}(q',k)\,\hat{R}^{J1}_{+-}(k,q)], \\
\hat{R}^{J1}_{+-}(q',q) &= \hat{V}^{J1}_{+-}(q',q) + \mathcal{P}\int_0^\infty \mathrm{d}k\, k^2 \frac{M}{q^2-k^2}[\,\hat{V}^{J1}_{++}(q',k)\,\hat{R}^{J1}_{+-}(k,q) \\
&\quad + \hat{V}^{J1}_{+-}(q',k)\,\hat{R}^{J1}_{--}(k,q)], \\
\hat{R}^{J1}_{-+}(q',q) &= \hat{V}^{J1}_{-+}(q',q) + \mathcal{P}\int_0^\infty \mathrm{d}k\, k^2 \frac{M}{q^2-k^2}[\,\hat{V}^{J1}_{--}(q',k)\,\hat{R}^{J1}_{-+}(k,q) \\
&\quad + \hat{V}^{J1}_{-+}(q',k)\,\hat{R}^{J1}_{++}(k,q)],
\end{aligned} \tag{1.38}$$

where we have introduced the abbreviations $\hat{R}^{J1}_{++} \equiv \hat{R}^{J1}_{J+1,J+1}$, $\hat{R}^{J1}_{--} \equiv \hat{R}^{J1}_{J-1,J-1}$, $\hat{R}^{J1}_{+-} \equiv \hat{R}^{J1}_{J+1,J-1}$, $\hat{R}^{J1}_{-+} \equiv \hat{R}^{J1}_{J-1,J+1}$. Conventionally, the coupled triplet channels in NN scattering are considered in this form, which can also be obtained by decomposing (1.27) directly into LSJ states. In a non-relativistic consideration, it is the tensor force that couples triplet states with $L = J \pm 1$.

So far, we have never mentioned the total isospin of the two-nucleon system, T (which is either 0 or 1). The reason for this is simply that T is not an independent quantum number. That is, owing to the antisymmetry of the two-fermion state, the quantum numbers L, S, and T have to fulfill the condition

$$(-1)^{L+S+T} = -1\,. \tag{1.39}$$

Thus, for given L and S, T is fixed.

1.2.3 The On-Shell R-Matrix and Phase Shifts

Phase shifts are a parametrization of the unitary S-matrix, which for uncoupled cases is given by

$$S_J = \mathrm{e}^{2\mathrm{i}\delta_J}. \tag{1.40}$$

Using the above and (1.25) and (1.26) in partial-wave decomposition, one can relate the on-energy-shell R-matrix to thc phase shifts as follows:
Spin singlet

$$\tan\, {}^0\delta^J(E_{\rm lab}) = -\frac{\pi}{2} qM\, {}^0\hat{R}^J(q,q); \tag{1.41}$$

Uncoupled spin triplet

$$\tan\, {}^1\delta^J(E_{\rm lab}) = -\frac{\pi}{2} qM\, {}^1\hat{R}^J(q,q). \tag{1.42}$$

For the *coupled states*, a unitary transformation is needed to diagonalize the two-by-two coupled R-matrix. This requires an additional parameter, known as the "mixing parameter" ϵ_J. Using the convention introduced by Blatt and Biedenharn [1.19], the eigenphases for the coupled channels are, in terms of the on-shell $\hat{R}$-matrix,

$$\tan\, \delta^J_{\mp}(E_{\rm lab}) = -\frac{\pi}{4} qM \left(\hat{R}^J_{J-1,J-1} + \hat{R}^J_{J+1,J+1} \pm \frac{\hat{R}^J_{J-1,J-1} - \hat{R}^J_{J+1,J+1}}{\cos 2\epsilon_J} \right),$$

$$\tan 2\epsilon_J(E_{\rm lab}) = \frac{2\hat{R}^J_{J+1,J-1}}{\hat{R}^J_{J-1,J-1} - \hat{R}^J_{J+1,J+1}}\,. \tag{1.43}$$

All $\hat{R}$-matrix elements in these formulas carry the arguments (q,q) where q denotes the c.m. on-energy-shell momentum that is related to the energy in the laboratory system, $E_{\rm lab}$, by

$$E_{\rm lab} = 2q^2/M. \tag{1.44}$$

An alternative convention for the phase parameters has been used by Stapp et al. [1.20], known as "bar" phase shifts. These are related to the Blatt–Biedenharn parameters by

$$\begin{aligned}\bar{\delta}^J_+ + \bar{\delta}^J_- &= \delta^J_+ + \delta^J_-,\\ \sin(\bar{\delta}^J_- - \bar{\delta}^J_+) &= \tan 2\bar{\epsilon}_J/\tan 2\epsilon_J,\\ \sin(\delta^J_- - \delta^J_+) &= \sin 2\bar{\epsilon}_J/\sin 2\epsilon_J.\end{aligned} \tag{1.45}$$

Phase-shift output from our codes is stated in terms of the "bar" conventions.

Using the transformation (1.37), the phase parameters for the coupled case can also be expressed directly in terms of the helicity-state $\hat{R}$-matrix elements; one obtains

$$\begin{aligned}\tan\, \delta^J_\mp(E_{\text{lab}}) &= -\frac{\pi}{4}qM\\ &\times\left[{}^{12}\hat{R}^J + {}^{34}\hat{R}^J \mp \frac{{}^{12}\hat{R}^J - {}^{34}\hat{R}^J - 4\sqrt{J(J+1)}\,{}^{55}\hat{R}^J}{(2J+1)\cos 2\epsilon_J}\right],\\ \tan 2\epsilon_J(E_{\text{lab}}) &= -2\frac{\sqrt{J(J+1)}({}^{12}\hat{R}^J - {}^{34}\hat{R}^J) + {}^{55}\hat{R}^J}{{}^{12}\hat{R}^J - {}^{34}\hat{R}^J - 4\sqrt{J(J+1)}\,{}^{55}\hat{R}^J}\,.\end{aligned} \tag{1.46}$$

1.2.4 Effective Range Parameters

For low-energy S-wave scattering, $q\cot\delta$ can be expanded as a function of q:

$$\frac{q}{\tan\delta} = q\cot\delta \approx -\frac{1}{a} + \frac{1}{2}rq^2, \tag{1.47}$$

where a is called the scattering length and r the effective range.

Rewriting this equation for two different (small) on-shell momenta q_1 and q_2 (with $E^{(i)}_{\text{lab}} = 2q_i^2/M$ and phase shifts δ_i, $i = 1, 2$), we can determine the two unknown constants a and r:

$$r = \frac{4}{M}\,\frac{\frac{q_1}{\tan\delta_1} - \frac{q_2}{\tan\delta_2}}{E^{(1)}_{\text{lab}} - E^{(2)}_{\text{lab}}} \tag{1.48}$$

and

$$\frac{1}{a} = \frac{M}{4}\,r\,E^{(i)}_{\text{lab}} - \frac{q_i}{\tan\delta_i}, \tag{1.49}$$

with $i = 1$ or 2.

1.2.5 Using Thompson's Equation

The Thompson equation (1.21) is solved most conveniently in "check" notation (1.22–24). The corresponding equations for the $\check{R}$-matrix are obtained from (1.32–34) by replacing

$$\frac{M}{q^2 - k^2} \longmapsto \frac{1}{2E_q - 2E_k} = \frac{\frac{1}{2}(E_q + E_k)}{q^2 - k^2} \tag{1.50}$$

and

$$\hat{R} \longmapsto \check{R}. \tag{1.51}$$

This is easily understood by comparing (1.18) with (1.24). $\check{R}$ is defined in analogy to (1.22). The phase-shift relation is

$$\tan\, {}^0\delta^J(E_{\text{lab}}) = -\frac{\pi}{2} q E_q\, {}^0\check{R}^J(q, q) \tag{1.52}$$

and similarly for the other channels.

1.3 One-Boson-Exchange Potentials

1.3.1 Interaction Lagrangians and OBE Amplitudes

We use the following Lagrangians for meson–nucleon coupling:

$$\mathcal{L}_{\text{ps}} = -g_{\text{ps}} \bar{\psi} \mathrm{i} \gamma^5 \psi \varphi^{(\text{ps})}, \tag{1.53}$$

$$\mathcal{L}_{\text{pv}} = -\frac{f_{\text{ps}}}{m_{\text{ps}}} \bar{\psi} \gamma^5 \gamma^\mu \psi \partial_\mu \varphi^{(\text{ps})}, \tag{1.54}$$

$$\mathcal{L}_{\text{s}} = +g_{\text{s}} \bar{\psi} \psi \varphi^{(\text{s})}, \tag{1.55}$$

$$\mathcal{L}_{\text{v}} = -g_{\text{v}} \bar{\psi} \gamma^\mu \psi \varphi^{(\text{v})}_\mu - \frac{f_{\text{v}}}{4M} \bar{\psi} \sigma^{\mu\nu} \psi (\partial_\mu \varphi^{(\text{v})}_\nu - \partial_\nu \varphi^{(\text{v})}_\mu), \tag{1.56}$$

with ψ the nucleon and $\varphi^{(\alpha)}_{(\mu)}$ the meson fields (notation and conventions as in [1.12]). For isospin 1 mesons, $\varphi^{(\alpha)}$ is to be replaced by $\boldsymbol{\tau} \cdot \boldsymbol{\varphi}^{(\alpha)}$ with τ^l ($l = 1, 2, 3$) the usual Pauli matrices. ps, pv, s, and v denote pseudo-scalar, pseudo-vector, scalar, and vector coupling/fields, respectively.

The one-boson-exchange potential (OBEP) is defined as a sum of one-particle-exchange amplitudes of certain bosons with given mass and coupling. We use the six non-strange bosons with masses below 1 GeV/c^2. Thus,

$$V_{\text{OBEP}} = \sum_{\alpha=\pi,\eta,\rho,\omega,\delta,\sigma} V^{\text{OBE}}_\alpha, \tag{1.57}$$

with π and η pseudo-scalar, σ and δ scalar, and ρ and ω vector particles. The contributions from the iso-vector bosons π, δ, and ρ contain a factor $\boldsymbol{\tau}_1 \cdot \boldsymbol{\tau}_2$.

The above Lagrangians imply the following OBE amplitudes:[1]

[1]Strictly speaking, we give here the potential defined as i times the Feynman amplitude; furthermore, there is a factor of i for each vertex and propagator; since $\mathrm{i}^4 = 1$, we simply ignore these factors of i.

$$
\begin{aligned}
&\langle \boldsymbol{q}'\lambda_1'\lambda_2'|V_{\rm ps}^{\rm OBE}|\boldsymbol{q}\lambda_1\lambda_2\rangle \\
&\quad = -\frac{g_{\rm ps}^2}{(2\pi)^3}\bar{u}(\boldsymbol{q}',\lambda_1')\mathrm{i}\gamma^5 u(\boldsymbol{q},\lambda_1)\bar{u}(-\boldsymbol{q}',\lambda_2')\mathrm{i}\gamma^5 u(-\boldsymbol{q},\lambda_2) \\
&\quad\quad /[(\boldsymbol{q}'-\boldsymbol{q})^2+m_{\rm ps}^2]; \qquad (1.58)
\end{aligned}
$$

$$
\begin{aligned}
&\langle \boldsymbol{q}'\lambda_1'\lambda_2'|V_{\rm pv}^{\rm OBE}|\boldsymbol{q}\lambda_1\lambda_2\rangle \\
&\quad = \frac{1}{(2\pi)^3}\frac{f_{\rm ps}^2}{m_{\rm ps}^2}\bar{u}(\boldsymbol{q}',\lambda_1')\gamma^5 \cdot \\
&\quad\quad \gamma^\mu \mathrm{i}(q'-q)_\mu u(\boldsymbol{q},\lambda_1)\bar{u}(-\boldsymbol{q}',\lambda_2')\gamma^5\gamma^\mu \mathrm{i}(q'-q)_\mu u(-\boldsymbol{q},\lambda_2) \\
&\quad\quad /[(\boldsymbol{q}'-\boldsymbol{q})^2+m_{\rm ps}^2] \\
&\quad = \frac{f_{\rm ps}^2}{(2\pi)^3}\frac{4M^2}{m_{\rm ps}^2}\{\bar{u}(\boldsymbol{q}',\lambda_1')\gamma^5 u(\boldsymbol{q},\lambda_1)\bar{u}(-\boldsymbol{q}',\lambda_2')\gamma^5 u(-\boldsymbol{q},\lambda_2) \\
&\quad\quad +[(E'-E)/(2M)]^2\bar{u}(\boldsymbol{q}',\lambda_1')\gamma^5\gamma^0 u(\boldsymbol{q},\lambda_1)\bar{u}(-\boldsymbol{q}',\lambda_2')\gamma^5\gamma^0 u(-\boldsymbol{q},\lambda_2) \\
&\quad\quad +[(E'-E)/(2M)][\bar{u}(\boldsymbol{q}',\lambda_1')\gamma^5 u(\boldsymbol{q},\lambda_1)\bar{u}(-\boldsymbol{q}',\lambda_2')\gamma^5\gamma^0 u(-\boldsymbol{q},\lambda_2) \\
&\quad\quad +\bar{u}(\boldsymbol{q}',\lambda_1')\gamma^5\gamma^0 u(\boldsymbol{q},\lambda_1)\bar{u}(-\boldsymbol{q}',\lambda_2')\gamma^5 u(-\boldsymbol{q},\lambda_2)]\} \\
&\quad\quad /[(\boldsymbol{q}'-\boldsymbol{q})^2+m_{\rm ps}^2]; \qquad (1.59)
\end{aligned}
$$

$$
\begin{aligned}
&\langle \boldsymbol{q}'\lambda_1'\lambda_2'|V_{\rm s}^{\rm OBE}|\boldsymbol{q}\lambda_1\lambda_2\rangle \\
&\quad = -\frac{g_{\rm s}^2}{(2\pi)^3}\bar{u}(\boldsymbol{q}',\lambda_1')u(\boldsymbol{q},\lambda_1)\bar{u}(-\boldsymbol{q}',\lambda_2')u(-\boldsymbol{q},\lambda_2)/[(\boldsymbol{q}'-\boldsymbol{q})^2+m_{\rm s}^2]; \qquad (1.60)
\end{aligned}
$$

$$
\begin{aligned}
&\langle \boldsymbol{q}'\lambda_1'\lambda_2'|V_{\rm v}^{\rm OBE}|\boldsymbol{q}\lambda_1\lambda_2\rangle \\
&\quad = \frac{1}{(2\pi)^3}\Big[g_{\rm v}\bar{u}(\boldsymbol{q}',\lambda_1')\gamma_\mu u(\boldsymbol{q},\lambda_1) \\
&\quad\quad +\frac{f_{\rm v}}{2M}\bar{u}(\boldsymbol{q}',\lambda_1')\sigma_{\mu\nu}\mathrm{i}(q'-q)^\nu u(\boldsymbol{q},\lambda_1)\Big] \\
&\quad\quad \times\Big[g_{\rm v}\bar{u}(-\boldsymbol{q}',\lambda_2')\gamma^\mu u(-\boldsymbol{q},\lambda_2) \\
&\quad\quad -\frac{f_{\rm v}}{2M}\bar{u}(-\boldsymbol{q}',\lambda_2')\sigma^{\mu\nu}\mathrm{i}(q'-q)_\nu u(-\boldsymbol{q},\lambda_2)\Big] \\
&\quad\quad /[(\boldsymbol{q}'-\boldsymbol{q})^2+m_{\rm v}^2] \\
&\quad = \frac{1}{(2\pi)^3}\{(g_{\rm v}+f_{\rm v})\bar{u}(\boldsymbol{q}',\lambda_1')\gamma_\mu u(\boldsymbol{q},\lambda_1) \\
&\quad\quad -\frac{f_{\rm v}}{2M}\bar{u}(\boldsymbol{q}',\lambda_1')[(q'+q)_\mu+(E'-E)(g_\mu^0-\gamma_\mu\gamma^0)]u(\boldsymbol{q},\lambda_1)\} \\
&\quad\quad \times\{(g_{\rm v}+f_{\rm v})\bar{u}(-\boldsymbol{q}',\lambda_2')\gamma^\mu u(-\boldsymbol{q},\lambda_2) \\
&\quad\quad -\frac{f_{\rm v}}{2M}\bar{u}(-\boldsymbol{q}',\lambda_2')[(q'+q)_\mu+(E'-E)(g^{\mu 0}-\gamma^\mu\gamma^0)]u(-\boldsymbol{q},\lambda_2)\} \\
&\quad\quad /[(\boldsymbol{q}'-\boldsymbol{q})^2+m_{\rm v}^2]. \qquad (1.61)
\end{aligned}
$$

Working in the two-nucleon center-of-mass frame, the momenta of the two incoming (outgoing) nucleons are $\boldsymbol{q}$ and $-\boldsymbol{q}$ ($\boldsymbol{q}'$ and $-\boldsymbol{q}'$). $E \equiv \sqrt{M^2 + \boldsymbol{q}^2}$ and $E' \equiv \sqrt{M^2 + \boldsymbol{q}'^2}$. Using the BbS or Thompson equation, the four-momentum transfer between the two nucleons is $(q' - q) = (0, \boldsymbol{q}' - \boldsymbol{q})$. The Dirac equation is applied repeatedly in the evaluations for the pv-coupling, and the Gordon identity [1.12] is used in the case of the v-coupling. [Note that in (1.61), second line from the bottom, the term $(q' + q)_\mu$ carries μ as a *sub*script to ensure the correct sign of the space component of that term.] The propagator for vector bosons is

$$\mathrm{i}\frac{-g_{\mu\nu} + (q' - q)_\mu (q' - q)_\nu / m_{\mathrm{v}}^2}{-(\boldsymbol{q}' - \boldsymbol{q})^2 - m_{\mathrm{v}}^2}, \tag{1.62}$$

where we drop the $(q' - q)_\mu (q' - q)_\nu$-term, which vanishes on-shell, anyhow, since the nucleon current is conserved. The off-shell effect of this term was examined in [1.21] and was found to be unimportant.

The Dirac spinors in helicity representation are given by

$$u(\boldsymbol{q}, \lambda_1) = \sqrt{\frac{E + M}{2M}} \begin{pmatrix} 1 \\ \frac{2\lambda_1 q}{E+M} \end{pmatrix} |\lambda_1\rangle, \tag{1.63}$$

$$u(-\boldsymbol{q}, \lambda_2) = \sqrt{\frac{E + M}{2M}} \begin{pmatrix} 1 \\ \frac{2\lambda_2 q}{E+M} \end{pmatrix} |\lambda_2\rangle. \tag{1.64}$$

They are normalized covariantly, that is,

$$\bar{u}(\boldsymbol{q}, \lambda) u(\boldsymbol{q}, \lambda) = 1 \tag{1.65}$$

with $\bar{u} = u^\dagger \gamma^0$.

At each meson–nucleon vertex, a form factor is applied that has the analytical form

$$\mathcal{F}_\alpha[(\boldsymbol{q}' - \boldsymbol{q})^2] = \left[\frac{\Lambda_\alpha^2 - m_\alpha^2}{\Lambda_\alpha^2 + (\boldsymbol{q}' - \boldsymbol{q})^2}\right]^{n_\alpha}, \tag{1.66}$$

with m_α the mass of the meson involved, Λ_α the so-called cutoff mass, and n_α an exponent. Thus, the OBE amplitudes (1.58–61) are multiplied by $\mathcal{F}_\alpha^2$.

In practice, it is desirable to have the potential represented in partial waves, since scattering phase shifts are defined in such a representation and nuclear-structure calculations are conventionally performed in an LSJ basis. We will turn to this in the next subsection.

1.3.2 Partial-Wave Decomposition

The OBE amplitudes are decomposed into partial waves according to

$$\langle \lambda_1' \lambda_2' | V^J(q', q) | \lambda_1 \lambda_2 \rangle$$
$$= 2\pi \int_{-1}^{+1} \mathrm{d}(\cos\theta)\, d^J_{\lambda_1 - \lambda_2, \lambda_1' - \lambda_2'}(\theta) \langle \boldsymbol{q}' \lambda_1' \lambda_2' | V | \boldsymbol{q} \lambda_1 \lambda_2 \rangle, \tag{1.67}$$

where θ is the angle between $\boldsymbol{q}$ and $\boldsymbol{q}'$, and $d^J_{m,m'}(\theta)$ is the conventional reduced rotation matrix.

In the present code, the integration over $\cos\theta$ is performed numerically. The $d^J_{m,m'}(\theta)$ is expressed in terms of Legendre polynominals, $P_J(\cos\theta)$. The following types of integrals will occur repeatedly:

$$\begin{aligned}
I_J^{(0)} &\equiv \int_{-1}^{+1} \mathrm{d}t \, \frac{P_J(t)}{(\boldsymbol{q}'-\boldsymbol{q})^2+m_\alpha^2} \, \mathcal{F}_\alpha^2[(\boldsymbol{q}'-\boldsymbol{q})^2], \\
I_J^{(1)} &\equiv \int_{-1}^{+1} \mathrm{d}t \, \frac{tP_J(t)}{(\boldsymbol{q}'-\boldsymbol{q})^2+m_\alpha^2} \, \mathcal{F}_\alpha^2[(\boldsymbol{q}'-\boldsymbol{q})^2], \\
I_J^{(2)} &\equiv \frac{1}{J+1}\int_{-1}^{+1} \mathrm{d}t \, \frac{JtP_J(t)+P_{J-1}(t)}{(\boldsymbol{q}'-\boldsymbol{q})^2+m_\alpha^2} \, \mathcal{F}_\alpha^2[(\boldsymbol{q}'-\boldsymbol{q})^2], \\
I_J^{(3)} &\equiv \sqrt{\frac{J}{J+1}}\int_{-1}^{+1} \mathrm{d}t \, \frac{tP_J(t)-P_{J-1}(t)}{(\boldsymbol{q}'-\boldsymbol{q})^2+m_\alpha^2} \, \mathcal{F}_\alpha^2[(\boldsymbol{q}'-\boldsymbol{q})^2], \\
I_J^{(4)} &\equiv \int_{-1}^{+1} \mathrm{d}t \, \frac{t^2P_J(t)}{(\boldsymbol{q}'-\boldsymbol{q})^2+m_\alpha^2} \, \mathcal{F}_\alpha^2[(\boldsymbol{q}'-\boldsymbol{q})^2], \\
I_J^{(5)} &\equiv \frac{1}{J+1}\int_{-1}^{+1} \mathrm{d}t \, \frac{Jt^2P_J(t)+tP_{J-1}(t)}{(\boldsymbol{q}'-\boldsymbol{q})^2+m_\alpha^2} \, \mathcal{F}_\alpha^2[(\boldsymbol{q}'-\boldsymbol{q})^2], \\
I_J^{(6)} &\equiv \sqrt{\frac{J}{J+1}}\int_{-1}^{+1} \mathrm{d}t \, \frac{t^2P_J(t)-tP_{J-1}(t)}{(\boldsymbol{q}'-\boldsymbol{q})^2+m_\alpha^2} \, \mathcal{F}_\alpha^2[(\boldsymbol{q}'-\boldsymbol{q})^2], \qquad (1.68)
\end{aligned}$$

where $t \equiv \cos\theta$. Notice that these integrals are functions of q', q, m_α, Λ_α, and n_α.

We state the final expressions for the partial-wave OBE amplitudes in terms of the combinations of helicity amplitudes defined in (1.31). More details concerning their derivation are to be found in Appendix E of [1.6].

Pseudo-scalar bosons (η and π meson; for π apply an additional factor of $\boldsymbol{\tau}_1 \cdot \boldsymbol{\tau}_2$):
Pseudo-scalar coupling (ps)

$$\begin{aligned}
{}^0V^J_{\mathrm{ps}} &= C_{\mathrm{ps}}\,(F^{(0)}_{\mathrm{ps}}\,I^{(0)}_J + F^{(1)}_{\mathrm{ps}}\,I^{(1)}_J), \\
{}^1V^J_{\mathrm{ps}} &= C_{\mathrm{ps}}\,(-F^{(0)}_{\mathrm{ps}}\,I^{(0)}_J - F^{(1)}_{\mathrm{ps}}\,I^{(2)}_J), \\
{}^{12}V^J_{\mathrm{ps}} &= C_{\mathrm{ps}}\,(F^{(1)}_{\mathrm{ps}}\,I^{(0)}_J + F^{(0)}_{\mathrm{ps}}\,I^{(1)}_J), \\
{}^{34}V^J_{\mathrm{ps}} &= C_{\mathrm{ps}}\,(-F^{(1)}_{\mathrm{ps}}\,I^{(0)}_J - F^{(0)}_{\mathrm{ps}}\,I^{(2)}_J), \\
{}^{55}V^J_{\mathrm{ps}} &= C_{\mathrm{ps}}\,F^{(2)}_{\mathrm{ps}}\,I^{(3)}_J, \\
{}^{66}V^J_{\mathrm{ps}} &= -C_{\mathrm{ps}}\,F^{(2)}_{\mathrm{ps}}\,I^{(3)}_J, \qquad (1.69)
\end{aligned}$$

with

$$C_{\rm ps} = \frac{g_{\rm ps}^2}{4\pi}\,\frac{1}{2\pi M^2} \tag{1.70}$$

and

$$\begin{aligned} F_{\rm ps}^{(0)} &= E'E - M^2, \\ F_{\rm ps}^{(1)} &= -q'q, \\ F_{\rm ps}^{(2)} &= -M(E'-E). \end{aligned} \tag{1.71}$$

Alternatively, the pseudo-vector (pv) coupling can be used for pseudo-scalar mesons. The basic scheme is the same as above; just replace the subscript ps by pv and use

$$\begin{aligned} F_{\rm pv}^{(0)} &= E'E - M^2 + (E'-E)^2(E'E+3M^2)/(4M^2), \\ F_{\rm pv}^{(1)} &= -q'q + q'q(E'-E)^2/(4M^2), \\ F_{\rm pv}^{(2)} &= -(E'-E)[\tfrac{1}{4}(E'-E)^2 + E'E]/M, \end{aligned} \tag{1.72}$$

and

$$C_{\rm pv} = \frac{f_{\rm ps}^2}{4\pi}\,\frac{4M^2}{m_\alpha^2}\,\frac{1}{2\pi M^2}\,. \tag{1.73}$$

Defining

$$g_{\rm ps} = f_{\rm ps}\,\frac{2M}{m_{\rm ps}}\,, \tag{1.74}$$

we see that $C_{\rm pv} = C_{\rm ps}$.

Scalar coupling (s) (σ and δ boson; for δ apply an additional factor of $\boldsymbol{\tau}_1\cdot\boldsymbol{\tau}_2$):

$$\begin{aligned} {}^0V_{\rm s}^J &= C_{\rm s}\,(F_{\rm s}^{(0)}\,I_J^{(0)} + F_{\rm s}^{(1)}\,I_J^{(1)}), \\ {}^1V_{\rm s}^J &= C_{\rm s}\,(F_{\rm s}^{(0)}\,I_J^{(0)} + F_{\rm s}^{(1)}\,I_J^{(2)}), \\ {}^{12}V_{\rm s}^J &= C_{\rm s}\,(F_{\rm s}^{(1)}\,I_J^{(0)} + F_{\rm s}^{(0)}\,I_J^{(1)}), \\ {}^{34}V_{\rm s}^J &= C_{\rm s}\,(F_{\rm s}^{(1)}\,I_J^{(0)} + F_{\rm s}^{(0)}\,I_J^{(2)}), \\ {}^{55}V_{\rm s}^J &= C_{\rm s}\,F_{\rm s}^{(2)}\,I_J^{(3)}, \\ {}^{66}V_{\rm s}^J &= C_{\rm s}\,F_{\rm s}^{(2)}\,I_J^{(3)}, \end{aligned} \tag{1.75}$$

with

$$C_{\rm s} = \frac{g_{\rm s}^2}{4\pi}\,\frac{1}{2\pi M^2} \tag{1.76}$$

and

$$\begin{aligned}
F_{\mathrm{s}}^{(0)} &= -(E'E + M^2), \\
F_{\mathrm{s}}^{(1)} &= q'q, \\
F_{\mathrm{s}}^{(2)} &= M(E' + E).
\end{aligned} \tag{1.77}$$

Vector bosons (v) (ω and ρ meson; for ρ apply an additional factor of $\boldsymbol{\tau}_1 \cdot \boldsymbol{\tau}_2$):
Vector–vector coupling

$$\begin{aligned}
{}^{0}V_{\mathrm{vv}}^{J} &= C_{\mathrm{vv}}\,(2E'E - M^2)\, I_J^{(0)}, \\
{}^{1}V_{\mathrm{vv}}^{J} &= C_{\mathrm{vv}}\,(E'E\, I_J^{(0)} + q'q\, I_J^{(2)}), \\
{}^{12}V_{\mathrm{vv}}^{J} &= C_{\mathrm{vv}}\,(2q'q\, I_J^{(0)} + M^2\, I_J^{(1)}), \\
{}^{34}V_{\mathrm{vv}}^{J} &= C_{\mathrm{vv}}\,(q'q\, I_J^{(0)} + E'E\, I_J^{(2)}), \\
{}^{55}V_{\mathrm{vv}}^{J} &= -C_{\mathrm{vv}}\, ME\, I_J^{(3)}, \\
{}^{66}V_{\mathrm{vv}}^{J} &= -C_{\mathrm{vv}}\, ME'\, I_J^{(3)},
\end{aligned} \tag{1.78}$$

with

$$C_{\mathrm{vv}} = \frac{g_{\mathrm{v}}^2}{4\pi}\,\frac{1}{\pi M^2}\,. \tag{1.79}$$

Tensor–tensor coupling

$$\begin{aligned}
{}^{0}V_{\mathrm{tt}}^{J} &= C_{\mathrm{tt}}\,\{(q'^2 + q^2)(3E'E + M^2)\, I_J^{(0)} \\
&\quad + [q'^2 + q^2 - 2(3E'E + M^2)]q'q\, I_J^{(1)} - 2q'^2q^2\, I_J^{(4)}\}, \\
{}^{1}V_{\mathrm{tt}}^{J} &= C_{\mathrm{tt}}\,\{[4q'^2q^2 + (q'^2 + q^2)(E'E - M^2)]\, I_J^{(0)} \\
&\quad + 2(E'E + M^2)q'q\, I_J^{(1)} \\
&\quad - (q'^2 + q^2 + 4E'E)q'q\, I_J^{(2)} - 2q'^2q^2\, I_J^{(5)}\}, \\
{}^{12}V_{\mathrm{tt}}^{J} &= C_{\mathrm{tt}}\,\{[4M^2 - 3(q'^2 + q^2)]q'q\, I_J^{(0)} \\
&\quad + [6q'^2q^2 - (q'^2 + q^2)(E'E + 3M^2)]\, I_J^{(1)} + 2(E'E + M^2)q'q\, I_J^{(4)}\}, \\
{}^{34}V_{\mathrm{tt}}^{J} &= C_{\mathrm{tt}}\,\{-(q'^2 + q^2 + 4E'E)q'q\, I_J^{(0)} - 2q'^2q^2\, I_J^{(1)} \\
&\quad + [4q'^2q^2 + (q'^2 + q^2)(E'E - M^2)]\, I_J^{(2)} + 2(E'E + M^2)q'q\, I_J^{(5)}\}, \\
{}^{55}V_{\mathrm{tt}}^{J} &= C_{\mathrm{tt}}\, M\{[E'(q'^2 + q^2) + E(3q'^2 - q^2)]\, I_J^{(3)} - 2(E' + E)q'q\, I_J^{(6)}\}, \\
{}^{66}V_{\mathrm{tt}}^{J} &= C_{\mathrm{tt}}\, M\{[E(q'^2 + q^2) + E'(3q^2 - q'^2)]\, I_J^{(3)} - 2(E' + E)q'q\, I_J^{(6)}\},
\end{aligned} \tag{1.80}$$

with

$$C_{\mathrm{tt}} = \frac{f_{\mathrm{v}}^2}{4\pi}\,\frac{1}{8\pi M^4}. \tag{1.81}$$

Vector–tensor coupling

$${}^{0}V_{\mathrm{vt}}^{J} = C_{\mathrm{vt}}\, M[(q'^2 + q^2)\, I_J^{(0)} - 2q'q\, I_J^{(1)}],$$

$$
\begin{aligned}
{}^{1}V_{\rm vt}^{J} &= C_{\rm vt}\, M[-(q'^2+q^2)\, I_J^{(0)} + 2q'q\, I_J^{(2)}],\\
{}^{12}V_{\rm vt}^{J} &= C_{\rm vt}\, M[6q'q\, I_J^{(0)} - 3(q'^2+q^2)\, I_J^{(1)}],\\
{}^{34}V_{\rm vt}^{J} &= C_{\rm vt}\, M[2q'q\, I_J^{(0)} - (q'^2+q^2)\, I_J^{(2)}],\\
{}^{55}V_{\rm vt}^{J} &= C_{\rm vt}\, (E'q^2+3Eq'^2)\, I_J^{(3)},\\
{}^{66}V_{\rm vt}^{J} &= C_{\rm vt}\, (Eq'^2+3E'q^2)\, I_J^{(3)},
\end{aligned}
\tag{1.82}
$$

with

$$
C_{\rm vt} = \frac{g_{\rm v} f_{\rm v}}{4\pi}\, \frac{1}{2\pi M^3}. \tag{1.83}
$$

We use units such that $\hbar = c = 1$. Energies, masses, and momenta are in MeV. The potential is in units of MeV^{-2}. The conversion factor is $\hbar c = 197.3286$ MeV fm. If the user wants to relate our units and conventions to those used in the work of other researchers, he/she should compare our (1.32) and our phase-shift relation (1.41) with the corresponding equations in other works.

1.3.3 Meson Parameters and Two-Nucleon Predictions

As an example of a relativistic momentum-space OBEP, we give here the potential presented in Section 4 of [1.7], which has become known as the Bonn B potential.[2] Its description of the NN data is excellent. This potential uses the pseudo-scalar coupling for π and η and is constructed within the framework of the BbS equation (1.15). The $\hat{V}$ version of this potential (1.17) is appropriate for application in non-relativistic nuclear-structure physics, since it satisfies (1.18).

In Table 1.1 we give the meson parameters. The predictions for the deuteron and low-energy scattering parameters are displayed in Table 1.2 together with the experimental values. Some phase shifts for neutron–proton scattering are listed in Table 1.3. Plots of the phase shifts predicted by Bonn B are shown in Fig. 5.10 of [1.7] together with the predictions by other NN models and results from phase-shift analyses.

1.4 Numerics and Codes

Two computer codes are supplied. The first (named `SPRING1` on the disk) computes the momentum-space OBEP described in Sect. 1.3. The second (file `SPRING2`) calculates the R-matrix by solving (1.32–34).

[2]It is customary to denote the OBE parametrizations of the Bonn full model [1.6] by the first letters of the alphabet, A, B, or C. See, for example, Appendix A of [1.7].

Table 1.1. Parameters for the Bonn B potential

Meson	J^P	I^G	$g_\alpha^2/4\pi$	f_α/g_α	m_α(MeV)	Λ_α(GeV)	$2n_\alpha$
π	0^-	1^-	14.4		138.03	1.7	2
η	0^-	0^+	3		548.8	1.5	2
ρ	1^-	1^+	0.9	6.1	769	1.85	4
ω	1^-	0^-	24.5	0	782.6	1.85	4
δ	0^+	1^-	2.488		983	2.0	2
σ^a	0^+	0^+	8.9437		550	1.9	2
			(18.3773)		(720)	(2.0)	(2)

J, P, I, and G denote spin, parity, isospin, and G-parity of the meson, respectively. Λ_α and n_α are the parameters for the form factor (1.66).
[a] The σ parameters given in parentheses apply to the $T = 0$ NN potential.

1.4.1 Momentum-Space OBEP

The subroutine BONN computes the momentum-space OBEP in terms of partial-wave helicity matrix elements according to the formulas (1.69–83). Though the code uses several subroutines, the user needs to call only BONN without worrying about the other computer programs contained in the package.

Nevertheless, we will briefly explain the purpose of each of those subroutines. Equations (1.69, 75, 78, 80, 82) obviously show a common pattern. This common structure of the mathematical expressions for the six OBE amplitudes is calculated by the subroutine OBSTR. The factors $F_\alpha^{(i)}$ (with $i = 0, 1, 2, \ldots,$), which are functions of q' and q only, are calculated in the major subroutine BONN itself, while the integrals $I_J^{(i)}$ (with $i = 0, \ldots, 6$) are computed in OBAI. The transfer of the Fs and Is to OBSTR proceeds via the COMMON /COB/. They are stored in the arrays F and AI with F$(i+1) = F_\alpha^{(i)}$ and AI$(i+1,$M$) = I_J^{(i)}$, where the index M runs over mesons using the same coupling.

Almost all communication between BONN and its OB-subroutines is done by means of the COMMON /COB/, which consists of many variables and arrays. The dimensions of the arrays are chosen generously such that extensions of the code will in general not require any extensions of this basic COMMON block.

OBAI computes the integrals (1.68) numerically using a mesh of Gauss points provided by the subroutine GSET. Up to 96 Gauss points can be ordered from GSET. The number of Gauss points necessary to compute the integrals with sufficient numerical accuracy depends on $q', q,$ and m_α and is calculated by OBAI before GSET is called. Via the input-parameter set for BONN, the user sets a lower and upper limit for the number of mesh points to be used (see below).

The meson propagator and the form factor, which are both functions of $\cos\theta$, are calculated in the subroutine OBAA and transferred to OBAI in the

Table 1.2. Deuteron and low-energy scattering parameters as predicted by the Bonn B potential (Theory) and from empirical sources (Experiment)

	Theory	Experiment[a]
Deuteron:		
Binding energy (MeV)	2.2246	2.224575(9)
D-state probability (%)	4.99	
Quadrupole moment (fm^2)	0.278 [b]	0.2860(15)
Asymptotic D/S-state ratio	0.0264	0.0264(12)
Root-mean-square radius (fm)	1.9688	1.9635(45)
Neutron–proton low-energy scattering (scattering length a, effective range r):		
$^1S_0$: $a_{\rm np}$ (fm)	–23.75	–23.748(10)
$r_{\rm np}$ (fm)	2.71	2.75(5)
$^3S_1$: $a_{\rm t}$ (fm)	5.424	5.424(4)
$r_{\rm t} = \rho(0,0)$ (fm)	1.761	1.759(5)

[a] The figures in parentheses after the experimental values give the one-standard-deviation uncertainties in the last digits. References to experiment can be found in [1.7].

[b] Without meson-exchange current contributions, which add 0.009(5) fm^2.

array **AA**. Subroutine **LEGP** provides the Legendre polynominals and is called by **OBAI**.

The helicity matrix elements for each boson exchange are calculated individually in **OBSTR** and stored separately in the array **VJ** of **COMMON /COB/**. These contributions from the individual mesons are finally added up at the end of the major subroutine **BONN** and stored in the array **V** of **COMMON /CPOT/**, which is used to transfer the potential values to the program that is calling **BONN**.

Input and output for the code system **BONN** are in units of MeV for momenta, masses, and cutoff masses, and in units of MeV^{-2} for the potential. However, *within* the code system **BONN**, all momenta, masses, and cutoff masses are in units of the nucleon mass M.

1.4.2 $\hat{R}$-Matrix and Phase Shifts

The $\hat{R}$-matrix is calculated by solving the integral equations (1.32–34). We apply the matrix-inversion method as introduced by Haftel and Tabakin [1.22].

To explain the method, let us consider an uncoupled case, for example, (1.32). First, we add to the right-hand side of this equation the term

$$-Mq^2\,{}^0\hat{V}^J(q',q)\,{}^0\hat{R}^J(q,q)\,\mathcal{P}\int_0^\infty \frac{\mathrm{d}k}{q^2-k^2}, \tag{1.84}$$

Table 1.3. Neutron–proton phase shifts (in degrees) in some low angular momentum partial waves for various laboratory energies (in MeV) as predicted by the Bonn B potential

State	25	50	100	200	300
$^1S_0$	50.72	39.98	25.19	5.66	−8.18
$^3P_0$	9.34	12.24	9.80	−1.02	11.48
$^1P_1$	−7.20	−11.15	−16.31	−23.47	−28.70
$^3P_1$	−5.33	−8.77	−13.47	−20.48	−26.38
$^3S_1$	80.32	62.16	41.99	19.03	4.07
$^3D_1$	−2.99	−6.86	−12.98	−20.29	−23.72
ϵ_1	1.76	2.00	2.24	3.03	4.03
$^1D_2$	0.68	1.58	3.34	6.21	7.49
$^3D_2$	3.88	9.29	17.67	24.94	25.36
$^3P_2$	2.62	6.14	11.74	16.65	17.40
$^3F_2$	0.11	0.34	0.77	1.10	0.52
ϵ_2	−0.86	−1.82	−2.84	−2.85	−2.02

which is equal to zero. This yields

$$\begin{aligned} {}^0\hat{R}^J(q',q) = {}^0\hat{V}^J(q',q) + \int_0^\infty \mathrm{d}k \, \frac{M}{q^2-k^2}[k^2\, {}^0\hat{V}^J(q',k)\, {}^0\hat{R}^J(k,q) \\ -q^2\, {}^0\hat{V}^J(q',q)\, {}^0\hat{R}^J(q,q)] \, , \end{aligned} \tag{1.85}$$

where we replaced the principal-value integral by an ordinary integral, since the added term makes the integrand smooth; for $k \to q$ the integrand is now finite.

The integral in (1.85) is computed by summing over N mesh points according to

$$\int_0^\infty \mathrm{d}k \, f(k) = \sum_{i=1}^{N} f(k_i) s_i \, . \tag{1.86}$$

We use Gaussian integration points and weights. Since the original Gauss points, x_i (with weights w_i), are in the interval $(-1,+1)$, we have to map them to the interval $(0,\infty)$, the range of our integration. For this purpose, we use the mapping

$$k_i = C \, \tan[\pi(x_i+1)/4], \tag{1.87}$$

with the new weights

$$s_i = C \, \frac{\pi}{4} \, \frac{w_i}{\cos^2[\pi(x_i+1)/4]}, \tag{1.88}$$

where C should be chosen in the order of 1000 MeV.

After rearranging (1.85),

$$ {}^0\hat{R}^J(q',q) - \int_0^\infty \mathrm{d}k \, \frac{M}{q^2-k^2}[k^2\, {}^0\hat{V}^J(q',k)\, {}^0\hat{R}^J(k,q) - q^2\, {}^0\hat{V}^J(q',q)\, {}^0\hat{R}^J(q,q)] $$
$$ = {}^0\hat{V}^J(q',q)\,, \tag{1.89} $$

we rewrite it in discrete form:

$$ \sum_{j=1}^{N+1} {}^0A^J_{ij}\, {}^0\hat{R}^J(k_j, k_{N+1}) = {}^0\hat{V}^J(k_i, k_{N+1}), \tag{1.90} $$

with $i = 1, \ldots, N+1$ and $k_{N+1} \equiv q$, the on-shell momentum. ${}^0A^J_{ij}$ is defined by

$$ {}^0A^J_{ij} = \delta_{ij} + u_j\, {}^0\hat{V}^J(k_i, k_j), \tag{1.91} $$

with

$$ u_j = Mk_j^2 \frac{s_j}{k_j^2 - q^2} \qquad \text{for } 1 \le j \le N \tag{1.92} $$

and

$$ u_{N+1} = -Mq^2 \sum_{j=1}^{N} \frac{s_j}{k_j^2 - q^2}, \tag{1.93} $$

where all grid points k_j have to be distinct from q. Notice that the "weight" u_{N+1} is just the added zero term that makes it possible to calculate the principal-value integral like an ordinary integral.

Defining ${}^0A^J \equiv ({}^0A^J_{ij})$, which is an $(N+1) \times (N+1)$ matrix, (1.90) can be written in matrix notation

$$ {}^0A^J\, {}^0\hat{R}^J = {}^0\hat{V}^J, \tag{1.94} $$

where ${}^0\hat{R}^J$ and ${}^0\hat{V}^J$ are vectors of dimension $N+1$. This is a system of $N+1$ linear equations with $N+1$ unknowns and has the familiar form

$$ AX = B, \tag{1.95} $$

with $X \equiv {}^0\hat{R}^J$ the unknown vector. It is solved by standard methods.

For the computation of the phase shifts, δ_l, only the on-shell $\hat{R}$-matrix element

$$ {}^0\hat{R}^J(q,q) \equiv {}^0\hat{R}^j(k_{N+1}, k_{N+1}) \tag{1.96} $$

is needed. From this, δ_l is obtained by means of (1.42). As a by-product, the half-off-shell $\hat{R}$-matrix, ${}^0\hat{R}^J(q',q) \equiv {}^0\hat{R}^J(k_i, k_{N+1})$, with $i = 1, \ldots, N$, is also obtained.

It is easy also to calculate the fully off-shell $\hat{R}$ matrix. Just consider ${}^0\hat{R}^J$ and ${}^0\hat{V}^J$ in (1.94) as $(N+1) \times (N+1)$ matrices with matrix elements ${}^0\hat{V}^J(k_i, k_j)$ and ${}^0\hat{R}^J(k_i, k_j)$ $(i, j = 1, \ldots, N+1)$. ${}^0A^J$ is the same as before. This option is built into the code (see below).

The extension to coupled channels is straightforward. In matrix notation, we have for the coupled case [cf. (1.34)]

$$\begin{pmatrix} {}^{12}A^J & {}^{55}\dot{A}^J \\ {}^{66}\dot{A}^J & {}^{34}A^J \end{pmatrix} \begin{pmatrix} {}^{12}\hat{R}^J & {}^{55}\hat{R}^J \\ {}^{66}\hat{R}^J & {}^{34}\hat{R}^J \end{pmatrix} = \begin{pmatrix} {}^{12}\hat{V}^J & {}^{55}\hat{V}^J \\ {}^{66}\hat{V}^J & {}^{34}\hat{V}^J \end{pmatrix}, \tag{1.97}$$

where the ${}^aA^J$ ($a = 12, 34, \ldots,$) are defined similarly to (1.91) and ${}^a\dot{A}^J \equiv {}^aA^J - \mathbf{1}$ with $\mathbf{1}$ the unit matrix. The matrix involving the As is a $(2N+2) \times (2N+2)$ matrix. If only the on-shell and half-off-shell $\hat{R}$ matrix is to be calculated, the matrices involving the $\hat{R}$s and $\hat{V}$s have two columns and $2N+2$ rows. If the fully off-shell $\hat{R}$ matrix is computed, all matrices in (1.97) are $(2N+2) \times (2N+2)$.

If $\hat{V}$ and $\hat{R}$ are considered in LSJ representation, the coupled system is [cf. (1.38)]

$$\begin{pmatrix} A^{J1}_{++} & \dot{A}^{J1}_{+-} \\ \dot{A}^{J1}_{-+} & A^{J1}_{--} \end{pmatrix} \begin{pmatrix} \hat{R}^{J1}_{++} & \hat{R}^{J1}_{+-} \\ \hat{R}^{J1}_{-+} & \hat{R}^{J1}_{--} \end{pmatrix} = \begin{pmatrix} \hat{V}^{J1}_{++} & \hat{V}^{J1}_{+-} \\ \hat{V}^{J1}_{-+} & \hat{V}^{J1}_{--} \end{pmatrix}. \tag{1.98}$$

For the coupled channels, our code offers both options, (1.97) and (1.98).

When building up the matrices in (1.97) or (1.98), our code takes advantage of the symmetries of $\hat{V}$: ${}^{12}\hat{V}^J$ and ${}^{34}\hat{V}^J$ are symmetric in q' and q, and ${}^{55}\hat{V}^J(q',q) = {}^{66}\hat{V}^J(q,q')$; in LSJ representation: $\hat{V}^{J1}_{++}$ and $\hat{V}^{J1}_{--}$ are symmetric in q' and q, and $\hat{V}^{J1}_{+-}(q',q) = \hat{V}^{J1}_{-+}(q,q')$.

For the coupled cases, the phase shifts are obtained from the on-shell $\hat{R}$-matrix elements by means of (1.43) or (1.46). Notice that for $\hat{R}^J_{--} \gg \hat{R}^J_{++}$ and ϵ_J small, the calculation of δ^J_+ may be inaccurate. This happens for the ${}^3S_1/{}^3D_1$ phase shifts around $E_{\text{lab}} \approx 16 - 18$ MeV, where $\delta^1_- \approx 90^0$. If more accuracy is needed for this particular case, an expansion of $\cos 2\epsilon_J$ should be used in (1.43).

The subroutine PHASES computes the $\hat{R}$ matrix and the phase shifts. Since it is a subroutine, there is a very small main program, CPH, that calls PHASES and, in addition, defines the potential subroutine to be applied and the dimensions of the various matrices and vectors needed in the computations of PHASES. For both vectors and matrices, vector arrays are used. Matrices are stored columnwise. The required dimensions depend on the number of Gauss points used and are specified in CPH by a parameter statement.

The notation used in PHASES is mostly identical to the one used in this section. Some differences in the notation are

$$\begin{aligned} \texttt{Q}(i) &= k_i, \\ \texttt{QO(K)} &= k_{N+1} = q, \\ \texttt{UQO} &= u_{N+1}, \end{aligned} \tag{1.99}$$

where K is an index that refers to the laboratory energy under consideration.

For the matrices and vectors involved in the system of linear equations, the notation of (1.95) is used in the code, where $B = \hat{V}^J$ *before* the solution, and $B = \hat{R}^J$ *after* the solution.

1.5 Using the Codes

A common feature of all codes is that most variables are transferred in common blocks. All routines that contain `READ` or `WRITE` statements have the `COMMON /CRDWRT/ KREAD,KWRITE,...`. The integer variable `KREAD` carries the number of the unit from which the input parameters are read. `KWRITE` is the corresponding unit number for output. The user should include the `COMMON /CRDWRT/ KREAD,KWRITE,...`, in his/her main program and define the input and output unit numbers. Input and output files carry names defined in open statements that appear in the beginning of the corresponding code.

1.5.1 OBEP Code `BONN`

`BONN` and all subroutines needed to run the code are contained in one file together with a short calling (main) program and a sample input and output file. To calculate the OBEP, only the subroutine `BONN` is called. At a single call, `BONN` computes six OBEP matrix elements for a given pair of momentum arguments (q', q) and a given total angular momentum J. When called for the first time, it also reads an input-parameter set from unit `KREAD`. As an example, the input-parameter set for the Bonn B potential (Sect. 1.3.3 and Table 1.1) is provided. The input is read in terms of fixed formats. Thus, if the user applies other meson parameter sets, he/she has to follow the formats of the sample set strictly. However, the number of mesons to be exchanged, their properties, the type of their coupling, and the cutoff parameters are variable.

The subroutine that does the reading and storing of the input parameters is `OBPAR`. It is called by `BONN` only once (the first time `BONN` is activated). We discuss now the meaning of the various records of the input file. The first record is a 70-character field which may be used to identify the potential. It is passed to the output file. All following records start with a 10-character field followed by either integer or real number(s). In the case of records 2 to 4, the 10-character field is used to name the parameter(s) for which values are given (on the record). The second record, starting with the identifier `FACTOR TYP`, provides in column 13 an integer parameter that can be `0, 1, or 2` and is stored in the array element `IFT(1)`. It tells the code by which factors the invariant version of the potential V as given in (1.69–83) is to be multiplied. For `IFT(1)=0`, no additional factors are applied, that is, following our notation introduced in Sect. 1.2.1, the invariant potential V is computed. If `IFT(1)=1`, the square-root factors of (1.16) are applied, that is, $\hat{V}$ is calculated. Finally, if `IFT(1)=2`, the factors of (1.23) are applied, so $\check{V}$ is provided. The third input record, starting with `NUM. INT.`, defines the minimum and maximum number of Gauss points to be used in the numerical angular integrations of (1.68). We recommend `4 48`. For test purposes, these parameters can be increased up to `96 96`. The fourth record, `NUCL. MASS`, gives the mass of the nucleon, which we take to be 938.926 MeV, the average between proton and neutron masses, appropriate for neutron–proton scattering.

The following records define the various mesons to be exchanged, their parameters, and cutoff parameters. By comparing Table 1.1 with the sample input file, it is easy to see how this part of the input has to be arranged. The first four characters of each record indicate either meson or cutoff parameters. If the record starts with `CUT`, it contains the cutoff parameters for the meson of the previous record. The first numerical parameter on a cutoff record identifies the type of cutoff to be used with `2.`, denoting the cutoff (1.66). Meson parameter records always start with the type of coupling to be applied, where the following abbreviations are used:

`0-` = pseudo-scalar,
`0-T` = pseudo-vector,
`0+` = scalar,
`1-T` = vector (including tensor) (1.100)

(cf. Sect. 1.3.1).

In general, all meson parameters are, of course, the same for the T=1 and the T=0 NN potential. However, in some OBE models (such as Bonn B), the σ boson is different for the two NN isospin states. In this case, the σ parameters and its cutoff parameters to be used for the $T = 0$ potential must be the last two records (before `END MESONS`) and must be preceded by a record that consists of between 4 and 10 asterisks (`**********`) and nothing else. The σ-parameters to be used for the T=1 potential must be right above the asterisks record. The records for the other mesons have to appear before all σ-records; otherwise, their order is arbitrary.

The last input record has to start with `END`.[3]

The transfer of arguments and values between the calling program and the potential subroutine `BONN` proceeds via two common blocks:

```
COMMON /CPOT/   V(6),XMEV,YMEV
COMMON /CSTATE/ J,HEFORM,SING,TRIP,COUP,ENDEP,LABEL
```

The user may ignore the logical variable `ENDEP` and the integer variable `LABEL`. We recommend setting, once and forever, in the calling program the logical variables `SING`, `TRIP`, and `COUP` to `.TRUE.`. Thus, only four parameters and `V` are of relevance. The calling program has to define `J`, the total angular momentum, and the relative momenta `XMEV` and `YMEV`, which are q' and q, respectively, in units of MeV. Given these arguments, the subroutine `BONN` computes six potential matrix elements and stores them in the array `V`. If `HEFORM=.TRUE.`, these six matrix elements are given in the helicity formalism;

[3]In the notation used in this chapter, quotations from the codes and their input/output are printed in upper case typewriter type style. Notice, however, that if the source codes are applied in lower case (as in the files provided), input control words like `END` or `1-T` have to be in lower case, that is, `end`, `1-t`, etc.

they are

$$
\begin{aligned}
\texttt{V(1)} &= {}^{0}V^{J}(q',q),\\
\texttt{V(2)} &= {}^{1}V^{J}(q',q),\\
\texttt{V(3)} &= {}^{12}V^{J}(q',q),\\
\texttt{V(4)} &= {}^{34}V^{J}(q',q),\\
\texttt{V(5)} &= {}^{55}V^{J}(q',q),\\
\texttt{V(6)} &= {}^{66}V^{J}(q',q)\,.
\end{aligned}
\tag{1.101}
$$

If HEFORM=.FALSE., the six matrix elements are given in the LSJ formalism; they are

$$
\begin{aligned}
\texttt{V(1)} &= V^{J0}_{J,J}(q',q),\\
\texttt{V(2)} &= V^{J1}_{J,J}(q',q),\\
\texttt{V(3)} &= V^{J1}_{J+1,J+1}(q',q),\\
\texttt{V(4)} &= V^{J1}_{J-1,J-1}(q',q),\\
\texttt{V(5)} &= V^{J1}_{J+1,J-1}(q',q),\\
\texttt{V(6)} &= V^{J1}_{J-1,J+1}(q',q)\,.
\end{aligned}
\tag{1.102}
$$

(Similar formulas apply to $\hat{V}$ and $\check{V}$.) Thus, depending on what formalism is used in the application (LSJ or helicity), the logical parameter HEFORM should be set accordingly at the beginning of the user's application main program.

We provide a small main program that calls the subroutine BONN and writes the potential matrix elements to an output file that is also given.

1.5.2 Code PHASES

All routines needed to run PHASES are contained in the second file SPRING2 provided, except for GSET and BONN (together with its subroutines OBPAR, OBSTR, OBAI, OBAA, and LEGP) that are contained in the potential package SPRING1. The input for PHASES is read from unit KREAD. A sample input and output file is provided. The input file consists of two parts. The second part is just the input for BONN (see above). The first part contains the parameters that control PHASES. The meaning of each input parameter is explained in the source code PHASES in comment line(s) preceeding the corresponding read statement. Therefore, there is no need to discuss them all here. Each input record starts with a 10-character field that is used to name the parameters for which values are provided. The information on the 10-character field has the relevance of a comment for the user and not of keywords for the code.

For the coupled channels, the code offers the option of solving the system of equations either in the helicity or the LSJ formalism (input parameter IHEF).

In a typical run, the phase shifts are written to unit KWRITE and there is no other output. If the first two values for E_{lab} are below 2 MeV, the low-energy scattering parameters in S-waves (cf. Sect. 1.2.4) are calculated and printed automatically.

In the sample run provided, the phase shifts of Table 1.3 and the low-energy scattering parameters in Table 1.2 are calculated.

If the $\hat{R}$ matrix is also to be written on a file, the integer input parameter IRMA must be non-zero (for example, 1). The $\hat{R}$ matrix is then written to the unit with unit number KPUNCH to be specified in the main program that calls PHASES (such a sample main program is included). If the fully off-shell $\hat{R}$ matrix is wanted, the parameter IQUA must also be non-zero. Note that the $\hat{R}$-matrix is always written in LSJ representation, no matter if the calculations were performed in the helicity or LSJ formalism. Because of the symmetries of $\hat{R}^{J1}_{L',L}(k_i,k_j)$, $i,j=1,\dots,N$, only the lower triangle of this matrix is written columnwise for each $\hat{R}^{J1}_{L',L}$.

We have discussed the formalism for the computation of the R matrix using the "hat" notation ($\hat{R}$), that is, we have considered the BbS equation in the form (1.18), which is identical to the non-relativistic Lippmann–Schwinger equation. We are, however, not restricted to this equation and $\hat{R}$. PHASES can also solve the Thompson equation. This is done in the "check" notation, as discussed in Sect. 1.2.5. The input parameters to be used for the two different scattering equations are the following:

		LS eq.[a]	Th. eq.[b]
in the input for BONN:	FACTOR TYP	1	2
in the input for PHASES:	IPROP	1	2
	IPHREL	1	2

[a] Lippmann–Schwinger equation (=BbS equation in "hat" notation), (1.18).
[b] Thompson equation in "check" notation (1.24).

1.6 What Else?

The nuclear-matter Brueckner G-matrix can be calculated in the same way as the $\hat{R}$-matrix of free-space scattering. The paper by Haftel and Tabakin [1.22] contains all details and all necessary formulas for nuclear-matter Brueckner calculations. The nuclear-matter results for the Bonn B potential are given in Section 9 of [1.7] (see also Chapter 2 of this book).

Potentials constructed in the framework of the Thompson equation are given in Appendix A, Table A.2, of [1.7]. These potentials use the pseudo-vector coupling for π and η. They are very appropriate for nuclear-matter calculations that include the (relativistic) medium effects on the Dirac spinors representing the nucleons in nuclear matter. This relativistic Dirac–Brueckner approach is sketched in Section 10.5 of [1.7] and discussed in detail in [1.8] where more results can be found.

Finally, we mention that the code PHASES can be used for the calculation of phase shifts and the $\hat{R}$-matrix applying any NN potential such as the Reid [1.23] or Paris [1.5] potential. However, the NN potential has to be given in momentum space (such codes exist for Reid and Paris). If a potential is represented in the LSJ formalism (which is usually the case), PHASES must be applied with the option IHEF=0. Furthermore, when applying non-relativistic potentials, the parameter settings IPROP=1 and IPHREL=1 must be used.

Acknowledgments

This work was supported in part by the U.S. National Science Foundation (NSF) through the Idaho-EPSCoR Program under Grant No. RII-8902065, by the NSF Physics Program under Grant No. PHY-8911040, and by the NSF-San Diego Supercomputer Center.

References

[1.1] J. Chadwick, Proc. Roy. Soc. (London) A**136**, 692 (1932)
[1.2] H. Yukawa, Proc. Phys. Math. Soc. Japan **17**, 48 (1935)
[1.3] A. D. Jackson, D. O. Riska, and B. Verwest, Nucl. Phys. A**249**, 397 (1975); G. E. Brown and A. D. Jackson, *The Nucleon–Nucleon Interaction* (North-Holland, Amsterdam, 1976)
[1.4] R. Vinh Mau, in: *Mesons in Nuclei*, edited by M. Rho and D. H. Wilkinson (North-Holland, Amsterdam, 1979) Vol. I, p. 151
[1.5] M. Lacombe et al., Phys. Rev. C**21**, 861 (1980)
[1.6] R. Machleidt, K. Holinde, and C. Elster, Phys. Reports **149**, 1 (1987)
[1.7] R. Machleidt, Adv. Nucl. Phys. **19**, 189 (1989)
[1.8] R. Brockmann and R. Machleidt, Phys. Lett. **149B**, 283 (1984), Phys. Rev. C**42**, 1965 (1990); H. Müther, R. Machleidt, and R. Brockmann, Phys. Rev. C**42**, 1981 (1990)
[1.9] E. E. Salpeter and H. A. Bethe, Phys. Rev. **84**, 1232 (1951)
[1.10] F. Gross, Phys. Rev. C**26**, 2203 (1982)
[1.11] R. Woloshyn and A. D. Jackson, Nucl. Phys. B**64**, 269 (1973)
[1.12] C. Itzykson and J. B. Zuber, *Quantum Field Theory* (McGraw-Hill, New York, 1980)
[1.13] R. Blankenbecler and R. Sugar, Phys. Rev. **142**, 1051 (1966)
[1.14] J. Fleischer and J. A. Tjon, Phys. Rev. D**21**, 87 (1980)
[1.15] G. E. Brown, A. D. Jackson, and T. T. S. Kuo, Nucl. Phys. A**133**, 481 (1969)
[1.16] R. H. Thompson, Phys. Rev. D**1**, 110 (1970)
[1.17] M. Jacob and G. C. Wick, Ann. Phys. (N.Y.) **7**, 404 (1959)
[1.18] K. Erkelenz, R. Alzetta, and K. Holinde, Nucl. Phys. A**176**, 413 (1971)
[1.19] J. Blatt and L. Biedenharn, Phys. Rev. **86**, 399 (1952)
[1.20] H. P. Stapp, T. J. Ypsilantis, and N. Metropolis, Phys. Rev. **105**, 302 (1957)
[1.21] K. Holinde and R. Machleidt, Nucl. Phys. A**247**, 495 (1975)
[1.22] M. I. Haftel and F. Tabakin, Nucl. Phys. A**158**, 1 (1970)
[1.23] R. V. Reid, Ann. Phys. (N. Y.) **50**, 411 (1968)

2. The G-Matrix in Finite Nuclei

H. Müther and P.U. Sauer

2.1 Introduction

One of the central aims of nuclear structure theory is to derive the properties of nuclei from a realistic nucleon–nucleon (NN) interaction. Such a realistic NN interaction should be determined to describe the interaction of two nucleons in a vacuum, which means that it should reproduce the experimental data on NN scattering and the properties of the deuteron. Typical examples of such realistic NN interactions include purely phenomenological NN potentials, assuming a hard-core [2.1] or a soft-core [2.2] parameterization of the short-range repulsion, but also more microscopic approaches based on the meson-exchange model [2.3,4]. Assuming one of these realistic interactions, one tries to solve the many-body problem of A interacting nucleons and compares the results obtained for observables characterizing the ground-state and low-energy excitations with the experimental data.

The scheme just outlined is based on the traditional model of a nucleus being composed of nucleons as the elementary building blocks. Of course, it is now generally accepted that the nucleons themselves cannot be regarded as inert particles but contain a substructure of quarks and gluons. Such a composite nucleon, however, might be modified in the nuclear medium. The form factor or radius of a nucleon could be affected by the nuclear medium, but also the interaction of two nucleons inside a nucleus might be quite different from the interaction of two nucleons in a vacuum. It is one of the main aims of modern nuclear physics to find examples with clear evidence for the effects of such subnucleonic degrees of freedom in nuclear systems at low energies. Before one can draw any conclusion about the necessity of considering subnucleonic degrees of freedom to understand specific features observed in experiments, one has to investigate carefully if the same feature might not already be described within the traditional model. Therefore, detailed nuclear-structure investigations based on many-body calculations employing realistic interactions within the traditional model are required to explore the effects of subnucleonic degrees of freedom, which go beyond this model.

In such nuclear-structure investigations, one has to deal with the problems introduced by the strong components of short range contained in all realistic NN interactions. This problem is evident in particular when so-called hard-core potentials are employed, which are infinite for relative distances smaller than the radius of the hard core r_c. The matrix elements of such a potential V evaluated for an uncorrelated two-body wave function $\Phi(r)$ diverge, since $\Phi(r)$ is different from zero also for relative distances r smaller than the hard-core radius r_c (see the schematic Fig. 2.1). A way out of this problem is to account for the two-body correlations induced by the NN interaction in the

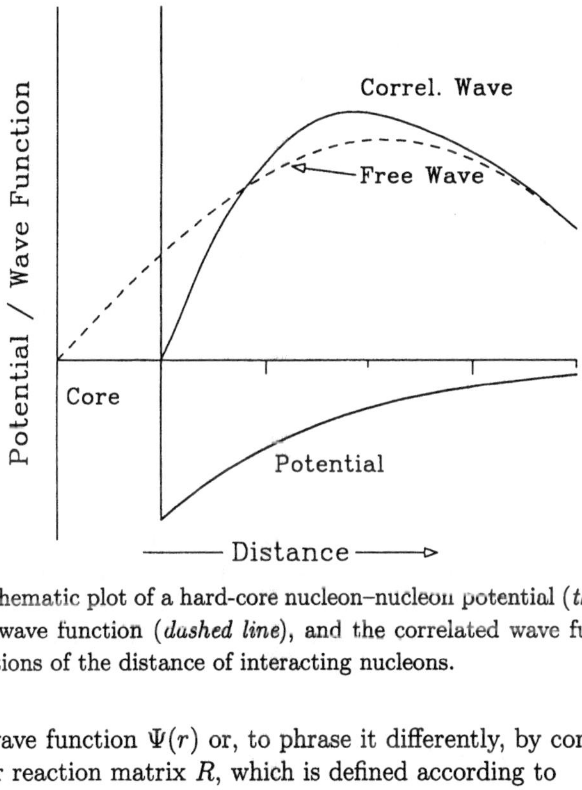

Fig. 2.1. Schematic plot of a hard-core nucleon–nucleon potential (*thick line*), the free relative wave function (*dashed line*), and the correlated wave function (*solid line*) as functions of the distance of interacting nucleons.

correlated wave function $\Psi(r)$ or, to phrase it differently, by considering the scattering or reaction matrix R, which is defined according to

$$\langle\Phi|R|\Phi\rangle = \langle\Phi|V|\Psi\rangle \ . \tag{2.1}$$

Since by definition the R-matrix effectively takes into account the correlations in the two-body wave function, its matrix elements between uncorrelated states $|\Phi\rangle$ will be finite even for a hard-core potential V (see Fig. 2.1). The correlations tend to enhance the amplitude of the correlated wave function Ψ relative to the uncorrelated one at distances r for which the interaction is attractive. A reduction of the amplitude is to be expected for small distances for which $V(r)$ is repulsive. This reduction turns into a complete suppression inside the core radius for a hard-core potential. From this discussion we see that the correlation effects tend to make the matrix elements of R more attractive than those of the bare potential V. For two nucleons in the vacuum, the R matrix can be determined by solving a kind of Lippmann–Schwinger equation:

$$R|\Phi\rangle = V\left\{|\Phi\rangle + \frac{1}{W - H_0 + \mathrm{i}\epsilon} V|\Psi\rangle\right\}$$
$$= \left\{V + V\frac{1}{W - H_0 + \mathrm{i}\epsilon}R\right\}|\Phi\rangle. \tag{2.2}$$

In these equations, the starting energy W stands for the energy of the two interacting nucleons, and H_0 represents the operator for the nucleons in the intermediate state without residual interaction, i.e., the kinetic energy. The G matrix may now be interpreted as the reaction matrix for two nucleons interacting in a nuclear medium. Therefore, in straight analogy to (2.2), we may write

$$G|\Phi\rangle = V|\tilde{\Psi}\rangle = V\left\{|\Phi\rangle + \frac{Q}{W - H_0 + \mathrm{i}\epsilon} V|\tilde{\Psi}\rangle\right\} \tag{2.3}$$

or the corresponding Bethe–Goldstone equation for the G operator

$$G = V + V\frac{Q}{W - H_0 + \mathrm{i}\epsilon}G. \tag{2.4}$$

In the nuclear medium, the Pauli principle has to be obeyed so that the difference between the correlated wave function $|\tilde{\Psi}\rangle$ in the medium and the uncorrelated one, which we will refer to as the defect function, can only contain components that are compatible with the Pauli principle. Therefore, the second term in the curly brackets on the right-hand side of (2.3), which represents the defect function, contains the Pauli operator Q, which does not allow for any components with one of the interacting nucleons scattered into states occupied by other nucleons:

$$\chi_{\text{Defect}} = \frac{Q}{W - H_0 + \mathrm{i}\epsilon} V|\tilde{\Psi}\rangle. \tag{2.5}$$

It is quite plausible that any restriction in the expansion of the defect function, as caused by the Pauli operator, tends to reduce the attraction obtained from the correlation effects. This effect, which will generally yield less attraction in the G matrix as more restrictions are imposed by the Pauli operator, is referred to as the Pauli quenching of the G matrix.

The surrounding medium furthermore influences the effective interaction G by means of the starting energy. For two nucleons bound in a nucleus, the starting energy will be negative. Therefore, the absolute values for the energy denominator in (2.4) will be larger the more the two interacting nucleons are bound, which also tends to reduce the attractive components of G due to the correlations. This effect is often referred to as the dispersive quenching of G. Negative values of W also prevent poles in the energy denominators of (2.3–5), so that the term $\mathrm{i}\epsilon$ becomes redundant in these equations. For such starting energies, no states exist in the expansion of the defect function (2.5) with the energy of the interacting pair of nucleons, and the defect function vanishes for large distances. This is known as the healing property of the defect function represented by the G matrix (see Fig. 2.1). The lack of poles also leads to real matrix elements of G.

From this discussion, we see that already the G matrix approximation to the effective interaction of two nucleons gives rise to a dependence of the interaction on the medium or density of the nuclear system. This density dependence, which is due to correlation effects within the traditional model of nuclear physics, must be well-understood before one can draw any firm conclusions about a medium dependence of the NN interaction owing to sub-nucleonic degrees of freedom (see discussion above).

It has already been mentioned that one must account for two-body correlations, or perform nuclear structure calculations in terms of the G matrix, if a hard-core potential [2.1] is employed, because in this case the matrix elements of the bare potential diverge. Essentially the same is also true if one considers so-called soft-core potentials like the one fitted to the NN scattering phase shifts by Reid [2.2]. If, using the Reid potential, the total energy of nuclear matter is calculated in Hartree–Fock approximation (which means ignoring two-body correlations), one obtains 176.2 MeV per nucleon at the empirical saturation density [2.5]. Taking into account the effects of NN correlations by performing such a calculation in terms of the G matrix, the so-called Brueckner–Hartree–Fock (BHF) approach, one obtains -10 MeV, which is much closer to the empirical value of -16 MeV per nucleon. Compared with this drastic change, caused by the consideration of two-nucleon correlations, the corrections to this key observable for realistic nuclear structure calculations are small though not negligible (an additional contribution of -5 MeV per nucleon has been evaluated from three-nucleon correlation [2.6]). Modern one-boson-exchange (OBE) potentials are typically softer in the sense that correlation effects are slightly smaller, but also for such potentials, the consideration of NN correlations is indispensable [2.5].

Most realistic nuclear structure calculations, including the evaluation of the G matrix, have been performed for infinite nuclear matter [2.5–7]. At first sight, the investigation of the infinite system seems to be easier than a direct evaluation of the properties of finite nuclei: the invariance of nuclear matter under translations determines the single-particle wave functions to be plane waves, whereas a Hartree–Fock type of calculation is needed for finite systems to produce a set of realistic single-particle functions. Such a general set of single-particle wave functions can lead to tremendous complications. As an example, we mention that the solution of the Bethe–Goldstone equation can be very complicated if the Pauli operator Q is defined in terms of such general functions.

For the discussion in this contribution and also for the computer program that calculates G, a set of oscillator functions is considered for which the oscillator parameter can be chosen in an appropriate way. More general wave functions could be expanded in a basis of such oscillator functions.

Being aware of these technical complications for realistic nuclear-structure calculations directly for finite nuclei, one can also collect good arguments why it may be easier to calculate in a reliable way the properties of finite nuclei than to develop a many-body theory to evaluate, for example, the saturation properties of nuclear matter: The effects of many-body correlations should increase with the average density of the system under consideration because the average number of nucleons so close together that they are within the range of the strong components for their mutual interaction increases with density. The typical nucleon densities for nuclei with small and medium mass numbers are considerably below the saturation density of nuclear matter. Therefore, a many-body calculation including effects up to A-nucleon correlations should give more accurate results for finite systems than for nuclear matter. Last but not least, however, what we observe in experiments are the properties of finite nuclei, and therefore one should perform structure calculations directly for these systems including all surface effects.

Following this introduction, Sect. 2.2 will give a different point of view of the G matrix as a first approximation for an effective NN interaction. General techniques and problems in solving the Bethe–Goldstone equation are discussed in Sect. 2.3, while Sect. 2.4 presents some details of the numerical methods used in a computer program. Some results for G in finite nuclei are presented and discussed in Sect. 2.5 with special emphasis on a comparison of different realistic NN interactions.

2.2 The G-Matrix, a First Step Towards an Effective NN Interaction

The solution of the Bethe–Goldstone equation (2.4) can be considered a first step in the Brueckner–Bethe–Goldstone theory (BBG) or hole-line expansion of the many-body theory for nuclear systems. In order to introduce the terminology for a short presentation of the BBG method, we write down the Schrödinger equation for the A-nucleon system,

$$\begin{aligned} H\Psi_i(1,\ldots,A) &= E_i\Psi_i(1,\ldots,A), \\ H &= T+V. \end{aligned} \tag{2.6}$$

In these equations, Ψ_i stands for the exact A-nucleon wave functions and E_i for the exact eigenvalues of the nuclear Hamiltonian H, which is split into the kinetic energy T and an operator for the NN interaction V. We also introduce an auxiliary single-particle potential U and rewrite H as

$$H = H_0 + H_1, \qquad H_0 = T + U, \qquad H_1 = V - U. \tag{2.7}$$

The single-particle potential U should be chosen such that H_1 is made small and can be treated as a perturbation. The A-nucleon eigenfunctions of H_0,

$$H_0\Phi_i(1,\ldots,A) = E_i^0\Phi_i(1,\ldots,A), \tag{2.8}$$

are used to define the model space. If one is interested in the ground-state properties of closed-shell nuclei like ^{16}O, this model space could be a single Slater determinant, for the case of low-energy states of open-shell nuclei, it could consist of those functions Φ_i that represent a typical set of basis states for a shell-model configuration-mixing calculation [2.8]. These functions Φ_i can therefore be used to define projection operators $\mathcal{P}$ and $\mathcal{Q}$, which project onto the model space and onto the rest of the A-nucleon Hilbert space, respectively. The aim is to derive an effective Hamiltonian H_{eff} such that

$$\mathcal{P}H_{\text{eff}}\mathcal{P}\Psi_i = E_i\mathcal{P}\Psi_i. \tag{2.9}$$

Several perturbation expansions have been formulated for this effective Hamiltonian. There are the energy-dependent expansions of Feshbach [2.9] and of Bloch and Horowitz [2.10], which correspond to the Brillouin–Wigner perturbation expansion and therefore yield an effective Hamiltonian that depends on the exact energy to be calculated. To get rid of this energy dependence in the case of multidimensional degenerate model spaces, one has to consider the so-called folded diagrams, which have been formulated, for example, by Brandow [2.11] and Kuo, Lee and Ratcliff [2.12]. Using this folded-diagram formulation, the effective Hamiltonian could be written as

$$H_{\text{eff}} = H_0 + H_1 + \left\{H_1\frac{\mathcal{Q}}{E^0 - H_0}H_{\text{eff}} + \text{folded diagrams}\right\}_{\text{linked}}. \tag{2.10}$$

The diagonalization of H_{eff} in the model space should yield the exact energies E_i and the projection of the exact wave function onto the model space $\mathcal{P}\Psi_i$. The terms in the curly brackets on the right-hand side of (2.10) represent the renormalization of the Hamiltonian to account for those degrees of freedom not explicitly considered in the model space. We do not want to go into details of the evaluation of such effective operators. Instead we would like to use this equation to summarize some statements about the appropriate use and limitations of the G-matrix approximation.

From the expansion in (2.10), it is obvious that H_{eff} will not only contain one-body and two-body operators but, in general, also operators involving three or more nucleons [2.12]. If, however, one ignores folded diagrams and only considers the so-called ladder diagrams (displayed in the upper half of Fig. 2.2) in the expansion of (2.10), the two-body part of H_{eff} is approximated by the G-matrix. In this case, the energy denominator reduces to the sum of the single-particle energies for the nucleons in the valence states minus the energies of the nucleons in the intermediate particle states, and corresponds to the denominator in the Bethe–Goldstone equation (2.4) if the energies for the particle states are approximated by the kinetic energy (see discussion below). In this case also, the projection operator $\mathcal{Q}$ of (2.10) is very similar to the Pauli operator of (2.4). Note, however, that, for example, for the model space

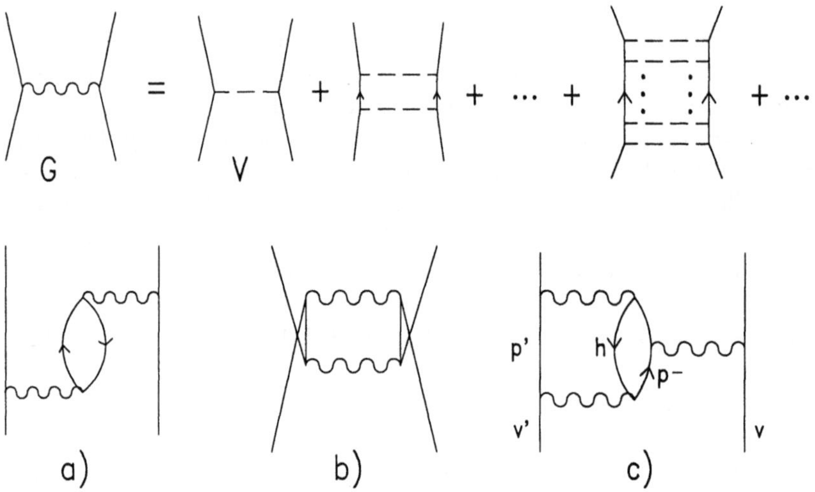

Fig. 2.2. The definition of the G-matrix (*wiggly line*) in terms of the ladder diagrams containing multiple V interactions is displayed in the upper half of the figure, whereas the lower part contains some diagrams representing terms of higher order in G, contributing to H_{eff} for the shell model. Lines with arrows pointing upwards denote valence – or particle – states, and hole lines are denoted by downward lines.

of a shell model, this Pauli operator would suppress intermediate states with one of the nucleons in a state occupied by the core nucleus and with both nucleons in the valence shell.

The two-body part of the effective operator to be used in shell-model calculations also contains terms beyond the G-matrix approximation like those displayed in the lower half of Fig. 2.2 [2.13]. Similar contributions also exist for other model spaces or nuclear-structure calculations. Such diagrams are frequently expressed in terms of G-matrix interactions, because a perturbation expansion in terms of V will not converge for realistic NN interactions. Note, however, that in general the choice of the starting energy to be used for the G-matrix elements not only depends on the single-particle energies of those states that are directly attached to the G-matrix interaction line but also involves the energies of many-body states [see (2.10)].

To illustrate this feature, we discuss as an example the starting energy to be used in the G-matrix element represented by the middle interaction line in diagram (c) of Fig. 2.2. This interaction involves nucleons in a valence state v and a particle state p (to simplify the notation, we consider the contribution with diagonal matrix elements for G). The on-shell choice for the starting energy would yield: $W = \epsilon_p + \epsilon_v$. According to the rules for this kind of Goldstone diagrams, however, the energy denominator for each intermediate state is the sum of energies for incoming valence lines plus the sum of single-particle energies for the hole lines minus the sum of single-particle energies of particle and valence lines present in that intermediate state [2.14]. In our example, this implies that the interaction between nucleons in states v and p should be calculated with a starting energy

$$W = \epsilon_v + \epsilon_{v'} + \epsilon_h - \epsilon_{p'}. \tag{2.11}$$

This off-shell definition for the starting energy demonstrates that the starting energy in general must be considered an independent dynamical variable of the G-matrix interaction.

The question about the appropriate choice for the starting energy must also be discussed for the calculation of ground-state properties for closed-shell nuclei. It is the aim to find an optimal single-particle potential U in the definition of the unperturbed Hamiltonian H_0 of (2.7). A choice frequently used is the Brueckner–Hartree–Fock (BHF) approximation:

$$\langle i|U|j\rangle = \begin{cases} \sum_{h\le F}\langle ih|\frac{1}{2}\left[G(W_{ih}) + G(W_{jh})\right]|jh\rangle & \text{if } i \text{ and } j \le F, \\ \sum_{h\le F}\langle ih|G(W_{ih})|jh\rangle & \text{if } i \le F \text{ and } j > F, \\ 0 & \text{if } i \text{ and } j > F, \end{cases} \tag{2.12}$$

with $W_{ab} = \epsilon_a + \epsilon_b$. For matrix elements of U involving hole states, it can be shown by a theorem of Bethe, Brandow, and Petschek (BBP) [2.15] that the on-shell definition of the starting energy in (2.12) yields an exact cancellation of many diagrams of higher order in G. The BBP theorem cannot be applied to the particle–particle matrix elements of U. The corresponding diagrams require an off-shell definition of the starting energy and therefore cannot be used to define a single-particle potential. The choice $U = 0$ presented in the last line of (2.12) may be justified in particular for the high-lying particle states for which the effects of a potential energy can be neglected compared with the effects of the kinetic energy. It is this so-called conventional choice for the particle spectrum that we have adopted in the Bethe–Goldstone equation (2.4) and that will be used throughout this chapter if not stated differently.

The optimal choice of H_0 to be used for the particle states in the BHF definition of U, and consequently in the Bethe–Goldstone equation, has widely been discussed, particularly for the case of nuclear-matter calculations [2.16,17]. Good arguments have been presented to favor a single-particle spectrum that is continuous at the Fermi surface and therefore contains an attractive potential for the low-lying particle states [2.17] or all particle states [2.16]. An answer to the problem of the optimal choice can only be obtained by evaluating the contributions of higher-order corrections and their sensitivity to the choice of U. Unfortunately, the binding energies calculated for nuclear matter using the BHF approximation show a significant dependence on U. The situation seems to be more stable for finite nuclei. In the case of ^{16}O, it has been demonstrated that the effects of three-body correlations could be minimized by assuming a constant potential for the particle states with a value as small as -8 MeV [2.18]. Such a constant shift can easily be incorporated into the solution of the Bethe–Goldstone equation by simply readjusting the starting energies (add -16 MeV). Also, with respect to the uncertainty for the optimal choice of U, the calculations for finite systems seemed to be more stable than those for nuclear matter (see also the discussion in the introduction). Having determined the BHF single-particle energies and wave functions $|h\rangle$ by diagonalizing the unperturbed Hamiltonian H_0 in a self-consistent procedure, the total energy is calculated in the BHF approximation by

$$E_{\mathrm{BHF}} = \frac{1}{2} \sum_{h \leq F} t_h + \epsilon_h \ , \tag{2.13}$$

the sum of single-particle energies ϵ_h and the expectation value of the kinetic energy t_h for the single-particle wave functions of the hole states.

2.3 The Solution of the Bethe–Goldstone Equation

As a first step towards determining the solution of the Bethe–Goldstone equation, we ignore the effects of the Pauli operator Q and evaluate the reaction matrix R of (2.2) for a given starting energy W. The corrections due to the Pauli operator will be considered in the next step. The technique used has been summarized in [2.19]. The reaction matrix R is obtained by solving the NN problem in momentum space and employing a matrix inversion technique very similar to the one discussed by Haftel and Tabakin [2.20]. The coordinates for the two interacting nucleons are represented by coordinates for the relative and center-of-mass motion. In coordinate space, the center-of-mass ($\boldsymbol{R}_{\mathrm{CM}}$) and relative coordinate ($\boldsymbol{r}$) are defined in terms of the coordinates ($\boldsymbol{r}_i$) of the interacting nucleons by

$$\boldsymbol{R}_{\mathrm{CM}} = \frac{1}{2}(\boldsymbol{r}_1 + \boldsymbol{r}_2) , \qquad \boldsymbol{r} = \boldsymbol{r}_1 - \boldsymbol{r}_2 , \tag{2.14}$$

and the corresponding relations for the momenta are given by

$$\hbar \boldsymbol{K} = \hbar \boldsymbol{k}_1 + \hbar \boldsymbol{k}_2, \qquad \hbar \boldsymbol{k} = \frac{1}{2}\left(\hbar \boldsymbol{k}_1 - \hbar \boldsymbol{k}_2\right). \tag{2.15}$$

For the moment we will assume that the motion of the center of mass is described by the wave function of a harmonic oscillator (oscillator energy spacing $\hbar\Omega$, quantum numbers N_{CM}, L_{CM}). The wave function for the relative motion is described in terms of plane waves and is expanded in a partial wave basis:

$$\begin{aligned} |klSJT\rangle &= i^l |kl\rangle |lSJT\rangle \\ &= \sum_{M_l, M_S} \langle lSM_lM_S|JM\rangle i^l Y_{lM_l}(\hat{k}) |kl\rangle |SM_S\rangle |TM_T\rangle \,. \end{aligned} \tag{2.16}$$

Here (S, M_S) and (T, M_T) refer to the coupled spin and isospin quantum numbers for the two interacting nuclei, Y_{lM_l} are the spherical harmonics for the orbital part of the relative wave function depending on the angular variables of the momentum vector $(\hat{k})$, and $\langle lSM_lM_S|JM\rangle$ refers to the Clebsch–Gordan coefficient coupling the orbital angular momentum l and the spin S to a total angular momentum J. The radial wave function $|kl\rangle$ shall be characterized by its coordinate representation,

$$\langle r|kl\rangle = \sqrt{\frac{2}{\pi}} j_l(kr), \tag{2.17}$$

with j_l the regular spherical Bessel functions. Since we are considering antisymmetrized two-nucleon states, we have to consider only those partial waves that are allowed by the Pauli principle, that is, states having $S + l + T$ equal to an odd integer. The partial-wave basis of (2.16) is very useful for defining the operator for a general NN interaction, which is symmetric with respect to rotation, parity, and isospin transformations. These symmetries imply that the interaction is diagonal with respect to the quantum numbers of total angular momentum, spin, and isospin. Therefore, the operator for such an interaction V can be decomposed as

$$\hat{V}(\boldsymbol{k}|\boldsymbol{k}') = \frac{2}{\pi}\frac{\hbar^2}{m} \sum_{J,S,T,l,l'} i^{l-l'} V_{ll'}^{JST}(k|k')\; |lJST\rangle\langle l'JST| \,. \tag{2.18}$$

Note that a tensor component in the NN interaction leads to matrix elements between partial-wave states of different orbital angular momenta l and l'. Since the total angular momentum and parity are conserved, this is possible only for triplet states ($S = 1$) with $l = J+1$, $l' = J-1$ or $l = J-1$, $l' = J+1$. A typical example for such coupled partial-wave channels is the $^3S_1$ – $^3D_1$ ($J = 1$, $S = 1$, $l, l' = 0, 2$, $T = 0$) states. The factor containing the nucleon mass m in front of the decomposition (2.18) is introduced only to shorten the notation in subsequent equations. Using this nomenclature, we can now rewrite the Lippmann–Schwinger equation [see (2.2)] for the R-matrix operator of free NN scattering in momentum representation:

$$\hat{R}(\boldsymbol{k}|\boldsymbol{k}') = \hat{V}(\boldsymbol{k}|\boldsymbol{k}') + \int \frac{\hat{V}(\boldsymbol{k}|\boldsymbol{q})\hat{R}(\boldsymbol{q}|\boldsymbol{k}')}{W - \frac{1}{2}\hbar\Omega\left(2N_{\mathrm{CM}} + L_{\mathrm{CM}} + \frac{3}{2}\right) - \frac{\hbar^2 q^2}{m}} \mathrm{d}q^3 \,. \quad (2.19)$$

Here the kinetic energy for the two nucleons in the intermediate states, which occurs in the denominator on the right-hand side of (2.18), has been expressed in terms of the kinetic energy for the center-of-mass motion (using the expectation value for the kinetic energy for the center-of-mass motion described by an oscillator function, see above) and for the relative motion with momentum $\hbar q$. Using the partial-wave decomposition for the potential given in (2.18) and a corresponding one for $\hat{R}$, the Lippmann–Schwinger equation (2.19) is separated into integral equations for the different partial waves (J, S, T) with an integral in one dimension, the absolute value for the relative momentum q:

$$R_{ll'}^{JST}(k|k') = V_{ll'}^{JST}(k|k') + \frac{2}{\pi}\sum_{\tilde{l}} \int_0^\infty \frac{V_{l\tilde{l}}^{JST}(k|q) R_{\tilde{l}l'}^{JST}(q|k')}{\tilde{W} - q^2} q^2 \,\mathrm{d}q \,. \quad (2.20)$$

The sum over intermediate orbital angular momenta $\tilde{l}$ is restricted for most partial-wave channels to $\tilde{l} = l = l'$. Only in the case of the tensor coupled channels does this sum also include $\tilde{l} = l + 2$ or $\tilde{l} = l - 2$. The quantity $\tilde{W}$ in the denominator of (2.20) is related to the starting energy and to the kinetic energy of the center-of-mass motion by

$$\tilde{W} = \frac{m}{\hbar^2}\left[W - \frac{1}{2}\hbar\Omega\left(2N_{\mathrm{CM}} + L_{\mathrm{CM}} + \frac{3}{2}\right)\right]. \quad (2.21)$$

In order to solve (2.20) numerically, the integral is approximated by introducing an appropriate N-point integration formula (see discussion in subsequent section)

$$\int_0^\infty f(k)\,\mathrm{d}k = \sum_{j=1}^{N} f(k_j)\,\omega_j \,. \quad (2.22)$$

Using this approximation for the numerical integration and suppressing the labels that refer to the orbital angular momenta and other partial-wave quantum numbers, the integral equation (2.20) is rewritten as

$$R(k_i|k_j) = V(k_i|k_j) + \frac{2}{\pi}\sum_{l=1}^{N} \frac{V(k_i|k_l)R(k_l|k_j)}{\tilde{W} - k_l^2} k_l^2\,\omega_l \,. \quad (2.23)$$

A simple algebraic manipulation brings this equation into the form

$$V(k_i|k_j) = \sum_{l=1}^{N} F(k_i|k_l)R(k_l|k_j), \quad (2.24)$$

with

$$F(k_i|k_l) = \delta_{il} - \frac{2}{\pi}\frac{V(k_i|k_l)}{\tilde{W} - k_l^2} k_l^2\,\omega_l \,, \quad (2.25)$$

which can be solved by matrix inversion techniques. For positive starting energies W, one may have to account for a singularity in the integral of (2.20) at $\tilde{W} = k_0^2$. For such positive energies, the diagonal amplitudes for the R matrix are related to the NN phase δ^{JST} in the case of uncoupled partial wave channels by [2.20]

$$\delta^{JST}(k_0) = \arctan\left[-k_0 R_{ll}^{JST}(k_0|k_0)\right] . \tag{2.26}$$

The amplitudes of R can also be used to evaluate matrix elements of the reaction matrix between oscillator wave functions U_{nl} for the relative motion

$$\begin{aligned}\langle nl|R(W;JSTN_{\mathrm{CM}}L_{\mathrm{CM}})|n'l'\rangle &= \frac{2}{\pi}\frac{\hbar^2}{m}\sum_{i,j=1}^{N} k_i^2 k_j^2 \omega_i\omega_j R_{ll'}^{JST}(k_i|k_j) \\ &\quad\times U_{nl}(k_i b_{rel}) U_{n'l'}(k_j b_{rel}) ,\end{aligned} \tag{2.27}$$

using the same mesh-points (2.22) for the integration of the matrix elements as were used for the solution of the Lippmann–Schwinger equation. The oscillator functions are normalized such that

$$\int_0^\infty U_{nl}(kb_{rel})U_{n'l}(kb_{rel})k^2\,\mathrm{d}k = \delta_{nn'} . \tag{2.28}$$

The oscillator parameter for the relative motion b_{rel} is defined by

$$b_{rel} = \sqrt{\frac{\hbar}{\mu\Omega}} = \sqrt{2}b, \tag{2.29}$$

with the reduced mass for the relative motion of two nucleons $\mu = m/2$ and b the corresponding oscillator parameter for single nucleons in the laboratory system. The oscillator matrix elements of the reaction matrix can then be used to account for the effects of the Pauli operator in the Bethe–Goldstone equation. For that purpose, we rewrite the Lippmann–Schwinger equation (2.2) and the Bethe–Goldstone equation (2.4) in an abbreviated form:

$$\begin{aligned} R &= V + VP_0R = V(1+P_0R) = V\mathcal{M}_0, \\ G &= V + VP_1G = V(1+P_1G) = V\mathcal{M}_1, \end{aligned} \tag{2.30}$$

with the propagators P_0 and P_1 for the free case and the Pauli corrected equation, respectively. Following the derivation in Appendix A of [2.15], we consider

$$G = G - R^\dagger(\mathcal{M}_1 - 1 - P_1G) + \left(\mathcal{M}_0^\dagger - 1 - R^\dagger P_0^\dagger\right)G , \tag{2.31}$$

where the expressions in parentheses on the right-hand side vanish by the definitions of the $\mathcal{M}_i$ and their Hermitian conjugates in (2.30). This can be transformed into

$$\begin{aligned} G &= R^\dagger - R^\dagger\mathcal{M}_1 + \mathcal{M}_0^\dagger G + R^\dagger\left(P_1 - P_0^\dagger\right)G \\ &= R^\dagger - \mathcal{M}_0^\dagger V^\dagger\mathcal{M}_1 + \mathcal{M}_0^\dagger V\mathcal{M}_1 + R^\dagger\left(P_1 - P_0^\dagger\right)G . \end{aligned} \tag{2.32}$$

If the potential is Hermitian ($V = V^\dagger$), the terms containing the wave operators $\mathcal{M}_i$ cancel in (2.32). If furthermore $P_i = P_i^\dagger$, the operators R and G are also Hermitian, and we can write

$$\begin{aligned} G &= R + R(P_1 - P_0)\, G \\ &= R + R\left(\frac{Q}{W - QTQ} - \frac{1}{W - T}\right) G\,. \end{aligned} \tag{2.33}$$

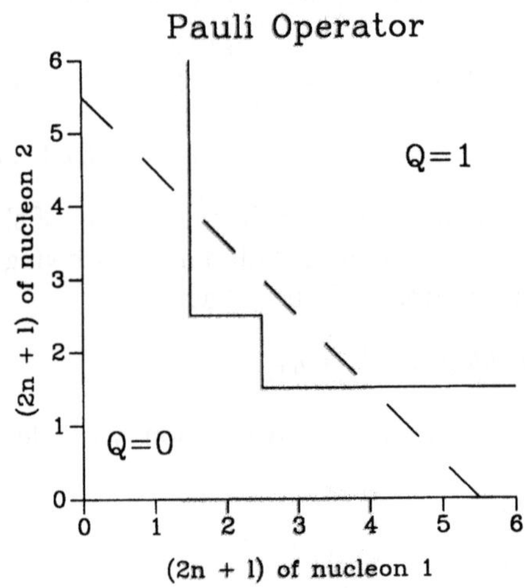

Fig. 2.3. The value for the Pauli operator as a function of the oscillator quantum numbers of the interacting nucleons. The domains for which $Q = 0$ and $Q = 1$ are separated by a continuous line for the example appropriate for the shell model of valence nucleons in the $1s0d$ shell outside an ^{16}O core. Also shown is the definition of a Pauli operator in the Eden–Emery approximation [*dashed line*, see (2.35)].

The Pauli operator is defined in terms of the "real" single-particle wave functions or those wave functions obtained, for example, in the BHF approximation, and a consistent treatment of Q can be rather complicated [2.21]. Even if we assume that the single-particle wave functions for the hole states can be approximated by harmonic-oscillator ones, the correct treatment of the Pauli operator is very troublesome, since Q is diagonal in the laboratory representation of the two-particle states only. Therefore, it is very common to approximate Q by an operator that is diagonal in the relative and center-of-mass representations. One example is the so-called angle-averaged Pauli operator [2.19,22]

$$\begin{aligned} Q(N_{\rm CM} L_{\rm CM} nl) &= \frac{1}{(2L_{\rm CM}+1)(2l+1)} \\ &\quad \times \sum_{\substack{n_1 l_1 n_2 l_2 \\ \lambda}} (2\lambda + 1)\, |\langle n_1 l_1 n_2 l_2 \lambda | N_{\rm CM} L_{\rm CM} nl \rangle|^2\,. \end{aligned} \tag{2.34}$$

The coefficients $\langle n_1 l_1 n_2 l_2 \lambda | N_{\rm CM} L_{\rm CM} n l \rangle$ are the transformation brackets [2.23] from the laboratory representation with two nucleons in harmonic-oscillator states defined by n_i and l_i to the harmonic-oscillator states in relative and center-of-mass coordinates. The summation over the single-particle quantum numbers (n_i, l_i) includes all states for which the Pauli operator is assumed to be one (see Fig. 2.3). Another example of a Pauli operator that is diagonal in the center-of-mass representation is the approximation introduced by Eden and Emery [2.24]:

$$Q_{\rm EE} = \begin{cases} 1, \text{ if } 2n_1 + l_1 + 2n_2 + l_2 = 2N_{\rm CM} + L_{\rm CM} + 2n + l \geq N_{\rm EE} \\ 0, \text{ if } 2n_1 + l_1 + 2n_2 + l_2 = 2N_{\rm CM} + L_{\rm CM} + 2n + l < N_{\rm EE} \end{cases}, \tag{2.35}$$

where $N_{\rm EE}$ can be chosen to obtain an optimal approximation (see Fig. 2.3). Employing such a Pauli operator and assuming in addition that the energy denominator is diagonal in the center-of-mass quantum numbers by using the approximation for the center-of-mass kinetic energy, $0.5\hbar\Omega(2N_{\rm CM}+L_{\rm CM}+3/2) = T_{N_{\rm CM}L_{\rm CM}}$, already considered in (2.19), the G-matrix elements are diagonal with respect to the center-of-mass quantum numbers, too. In this case, the Pauli corrections to the G-matrix can be determined easily by writing (2.33) in the oscillator basis for the relative motion:

$$\begin{aligned} \langle nl|G(W)|n'l'\rangle &= \langle nl|R(W)|n'l'\rangle + \sum_{n_1 l_1 n_2 l_2} \langle nl|R(W)|n_1 l_1\rangle \\ &\times \langle n_1 l_1 \left| Q \frac{1}{W - Q\left(T_{N_{\rm CM}L_{\rm CM}} + T_{rel}\right) Q} Q - \frac{1}{W - T_{N_{\rm CM}L_{\rm CM}} + T_{rel}} \right| n_2 l_2 \rangle \\ &\times \langle n_2 l_2 | G(W) | n'l'\rangle \, . \end{aligned} \tag{2.36}$$

To shorten the notation, we have suppressed all quantum numbers $(S, T, J, N_{\rm CM}, L_{\rm CM})$ that are diagonal in all the matrix elements considered in this equation. Employing an approximation for the Pauli operator as defined in (2.34), the property of a projection operator ($Q^2 = Q$) is lost. We have chosen to interpret the propagator $Q/(W - H_0)$ to mean $Q(W - H_0)^{-1}Q$ instead of using a non-symmetrical interpretation [2.19]. The G-matrix elements $\langle nl|G(W; S, T, J, N_{\rm CM}, L_{\rm CM})|n'l'\rangle$ are obtained by solving (2.36) with standard techniques for the solution of linear equations.

2.4 Details of the Numerical Calculation

This section contains some details of the numerical solution of the Bethe–Goldstone equation in a basis of harmonic-oscillator states for the relative and center-of-mass motion of the interacting nucleons. All references to subroutines refer to the FORTRAN program `CALGM`, which is distributed with this book or can be obtained from one of the authors (H.M.).

As has been discussed in the previous section, in our technique the first step towards the solution of the Bethe–Goldstone equation is to solve the free-scattering equation in momentum space (2.20). For this, one has to define a set of mesh points k_j and weight factors ω_j, which define the N-point integration formula (2.22) for the integral equation. This is done in the subroutine `GNMESH`, where the number of mesh points (NDIM) and the values for k_j and ω_j in fm^{-1} have to be specified. Without new specifications, the program defines NDIM=48 mesh points and corresponding weights by mapping the mesh points x_j for a Gauss–Legendre integration [2.25] from the interval $[-1,1]$ to the integration interval $[0,\infty]$ according to

$$k_j = \mathrm{SCALE}\ \tan\left[\frac{\pi}{4}(1+x_j)\right] , \tag{2.37}$$

using SCALE=1 fm^{-1}. For these meshpoints, one has to define the matrix elements $V_{ll'}^{JST}(k_i|k_j)$, which define the NN potential according to (2.18), separately for each partial-wave channel. As an example, we present these matrix elements for a central Yukawa potential assuming the exchange of a meson with mass $\tilde{\mu}$ and a coupling constant α (given in MeV). The matrix elements defining this potential in momentum space could be calculated as

$$\begin{aligned} V_{ll'}^{JST}(k_i|k_j) &= \delta_{ll'}\frac{\alpha m}{\hbar^2}\int_0^\infty r^2\,\mathrm{d}r\, j_l(k_ir)\frac{e^{-\tilde{\mu}r}}{r}j_l(k_jr) \\ &= \delta_{ll'}\frac{\alpha m}{\hbar^2}\frac{1}{k_ik_j}Q_l(z), \end{aligned} \tag{2.38}$$

with $z = (k_i^2+k_j^2+\tilde{m}^2)/2k_ik_j$ and $Q_l(z)$ the Legendre function of the second kind [2.25] defined by

$$\begin{aligned} Q_0(z) &= \frac{1}{2}\ln\left(\frac{z+1}{z-1}\right), \\ Q_1(z) &= \frac{z}{2}\ln\left(\frac{z+1}{z-1}\right)-1, \\ (l+1)Q_{l+1}(z)+lQ_{l-1}(z) &= (2l+1)zQ_l(z). \end{aligned} \tag{2.39}$$

In its present form, the program `CALGM` is set up to evaluate either the matrix elements for the Reid soft-core potential [2.2] (this is done by setting the input parameter IINTER to 1, using subroutine `GENPOT`) or by employing an OBE potential of the Bonn group [2.3] (for IINTER different from 1, the subroutine is used as an interface to the subroutine evaluating the OBE potential).

The Lippmann–Schwinger equation (2.20) is solved by matrix inversion in subroutine `GENTOS` for the partial-wave channel under consideration. The amplitudes $V_{ll'}^{JST}(k_i|k_j)$ and $R_{ll'}^{JST}(k_i|k_j)$ for this channel are stored on a disk file with unit number 8 in an unformatted direct-access file (see OPEN statement in the main program).

The oscillator matrix elements for R are calculated according to (2.27) in subroutine `TRFMAT` after the oscillator functions have been prepared in subroutines `GENGAM` and `GENOSC`. The oscillator functions, stored in the array

OSCARR for the mesh points $b_{\rm rel}k_j$ with k_j defined in (2.37), are all positive for large values of the argument. This implies that all matrix elements for the operator of the kinetic energies (also those non-diagonal in the oscillator quantum number) are positive. On the other hand, the corresponding oscillator functions in r space should all be positive for small values of r.

The direct integration of oscillator matrix elements according to (2.27) yields accurate results only if the number of mesh points is much larger than the number of nodes in the oscillator functions. For larger oscillator quantum numbers n, it may be more appropriate to evaluate the integral approximately by assuming that the integrand is different from zero only for momenta close to

$$\hat{k} = \frac{1}{b_{rel}}\sqrt{2(2n+l+\frac{3}{2})}\,. \tag{2.40}$$

The program in its present form employs this approximation for $n \geq 12$. This can be changed by modifying the value for the constant NMETH1 in subroutine `GNOSCD`.

Using the oscillator matrix elements for R, the Pauli corrected G-matrix elements are calculated in subroutine `CORR2` according to (2.36) and printed in the main subroutine `CALGM`. These oscillator matrix elements in the relative and center-of-mass representations could then be used to determine the antisymmetrized and normalized matrix elements of G in the laboratory representation coupled to total angular momentum $\mathcal{J}$ and isospin T by

$$\begin{aligned}
&\langle ab|G|cd\rangle_{\mathcal{J}T} = \\
&\quad \frac{1}{\sqrt{(1+\delta_{ab})(1+\delta_{cd})}}\sum_{\lambda\lambda' S}\left\{\begin{matrix} l_a & \frac{1}{2} & j_a \\ l_b & \frac{1}{2} & j_b \\ \lambda & S & \mathcal{J}\end{matrix}\right\}\left\{\begin{matrix} l_c & \frac{1}{2} & j_c \\ l_d & \frac{1}{2} & j_d \\ \lambda' & S & \mathcal{J}\end{matrix}\right\} \\
&\quad \times(-)^{\lambda+\lambda'}\sum_{nln'l'N_{\rm CM}L_{\rm CM}}\langle n_al_an_bl_b\lambda|N_{\rm CM}L_{\rm CM}nl\rangle \\
&\quad \times\langle n_cl_cn_dl_d\lambda'|N_{\rm CM}L_{\rm CM}n'l'\rangle(-)^{l+l'}\left[1-(-1)^{l+S+T}\right] \\
&\quad \times\sum_J(2J+1)\sqrt{(2\lambda+1)(2\lambda'+1)}W(L_{\rm CM}l\mathcal{J}S;\lambda J) \\
&\quad \times W(L_{\rm CM}l'\mathcal{J}S;\lambda' J)\langle nl|G(W;JSTN_{\rm CM}L_{\rm CM})|n'l'\rangle\,,
\end{aligned} \tag{2.41}$$

where we have used the conventional nomenclature to define the $9j$-symbols for the transformation j–j to L–S, the so-called Talmi–Moshinsky brackets [see also (2.34)] and the Racah coefficients W [2.8]. The quantum numbers l_a and j_a refer to orbital angular momentum and total angular momentum, respectively, for the nucleon in the single-particle orbital a.

2.5 Some Examples for Tests and Studies

At first sight, the two-step procedure for calculating the G-matrix as outlined above seems to introduce some unnecessary complications by solving

Table 2.1. Wound integral for various interactions. Results for the wound integral [$\int \chi^2 \mathrm{d}k$, see (2.42)] are listed for various partial waves and interactions. The quantum numbers in the first column can be identified from (2.42). All results are obtained for an uncorrelated oscillator function with $n' = 0$, b_{rel}=2.5 fm, and a starting energy $W = -90$ MeV. Results are listed for the Reid soft-core interaction [2.2] and the OBE potentials "A", "B", and "C" defined in Table A.1 of [2.3].

l	l'	S	J	Reid	OBEA	OBEB	OBEC
0	0	0	0	0.306	0.139	0.124	0.133
0	0	1	1	0.357	0.092	0.099	0.133
2	0	1	1	0.388	0.137	0.162	0.314

the integral equation for the free-scattering matrix R in momentum space first (2.20) and then, after transformation into an oscillator basis, accounting for the Pauli corrections by solving another set of linear equations (2.36). Would it not be more efficient to set up the Bethe–Goldstone equation with inclusion of the Pauli effects directly in the oscillator basis and solve only one set of linear equations? The answer to this question can be deduced easily from the results displayed in Fig. 2.4. There we show a defect function for the free-scattering case as a function of relative momentum [compare (2.5)]

$$\begin{aligned} \chi_{\text{Defect}}(k; ln'l'JSWN_{\text{CM}}L_{\text{CM}}) &= k\langle kl|\frac{1}{W-T}R(JSWN_{\text{CM}}L_{\text{CM}})|n'l'\rangle \\ &= \frac{k}{\tilde{W}-k^2}\frac{2}{\pi}\int_0^\infty k'2\mathrm{d}k'\, R_{ll'}^{JST}(k|k')U_{n'l'}(k'b_{rel})\,, \end{aligned} \tag{2.42}$$

with $\tilde{W}$ as defined in (2.21) and $U_{n'l'}$ the oscillator function for the relative motion. The example displayed in Fig. 2.4 (dashed line) has been obtained for the $^1S_0$ partial-wave channel employing the Reid soft-core potential and is compared with the oscillator function $U_{n'l'}$ ($n' = l' = 0$) also multiplied by the momentum k. From this figure, one can see that non-negligible contributions to the defect function are obtained up to very high momenta ($k \approx 8$ fm^{-1}). If one tried to account for such high-momentum components in an oscillator basis, one would have to take into account oscillator states at least up to $n \approx 100$ [see estimate in (2.40) considering $b_{rel} = 2.5$ fm]. This demonstrates that, in order to account for the short-range correlations of the NN interaction, it is more efficient to solve the scattering equation in momentum space. Since the Pauli correction [$(1-Q)$ in (2.36)] is significant for oscillator states with small quantum numbers only, the set of linear equations to account for these corrections can be well-approximated by considering only a small number (up to $n \approx 20$) in the summation of (2.36). Therefore, similar strategies, solving first for a reference scattering matrix and afterwards accounting for Pauli corrections, are also used in other methods to evaluate G for finite nuclei [2.26–28].

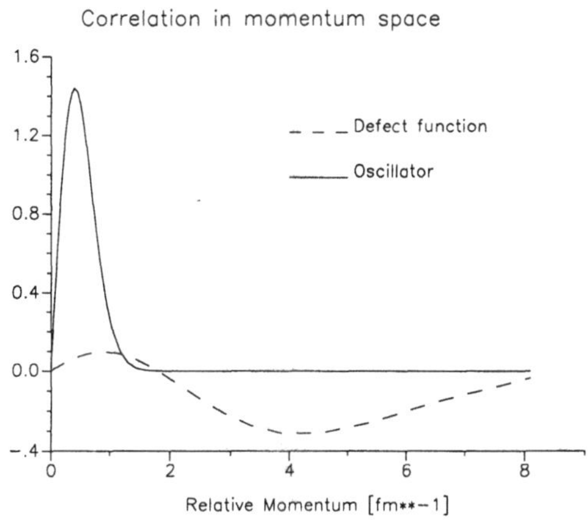

Fig. 2.4. The defect function in momentum space [dashed line, see definition in (2.42)] for the $^1S_0$ channel. The defect function has been calculated for the Reid potential assuming a starting energy $W = -90$ MeV for the oscillator state $n' = l' = N_{\rm CM} = L_{\rm CM} = 0$ and assuming an oscillator length $b_{\rm rel} = 2.5$ fm. This defect function is compared with the uncorrelated oscillator function for relative motion multiplied by the momentum k (solid line).

The defect function shown in Fig. 2.4 has been evaluated for the Reid potential. Similar results are obtained for other realistic NN potentials. The defect function is negative for large relative momenta (which after Fourier–Bessel transformation corresponds to small relative distances) and positive for small momenta (or larger distances). Therefore, it exhibits the same features as displayed in the schematic Fig. 2.1.

The integrated square of this defect function, often referred to as the wound integral, is a measure of the importance of correlation effects in a certain partial-wave channel. Results for the wound integral in partial waves with orbital angular momentum 0 are listed for the Reid potential and for three different OBE potentials [2.3] in Table 2.1. Comparing those results, one finds that the modern OBE potentials, in general, yield smaller values for the wound integral, which justifies the statement that these OBE potentials are "softer" than the Reid soft-core potential. The main difference between the three OBE potentials under consideration is the result for the D-wave component of the $^3S_1$ defect function [l=2, l'=0, in the nomenclature of Table 2.1 and (2.42)]. This value in particular characterizes the strength of the tensor components in the NN interaction. Therefore, these results reflect the reduction of the tensor component, comparing the Reid, OBEC, OBEB, and OBEA potential in a similar way as the D-state probabilities for the deuteron, which are calculated to be 6.5%, 5.61%, 4.99%, and 4.38%, respectively.

More details can be deduced from the comparison of the defect functions

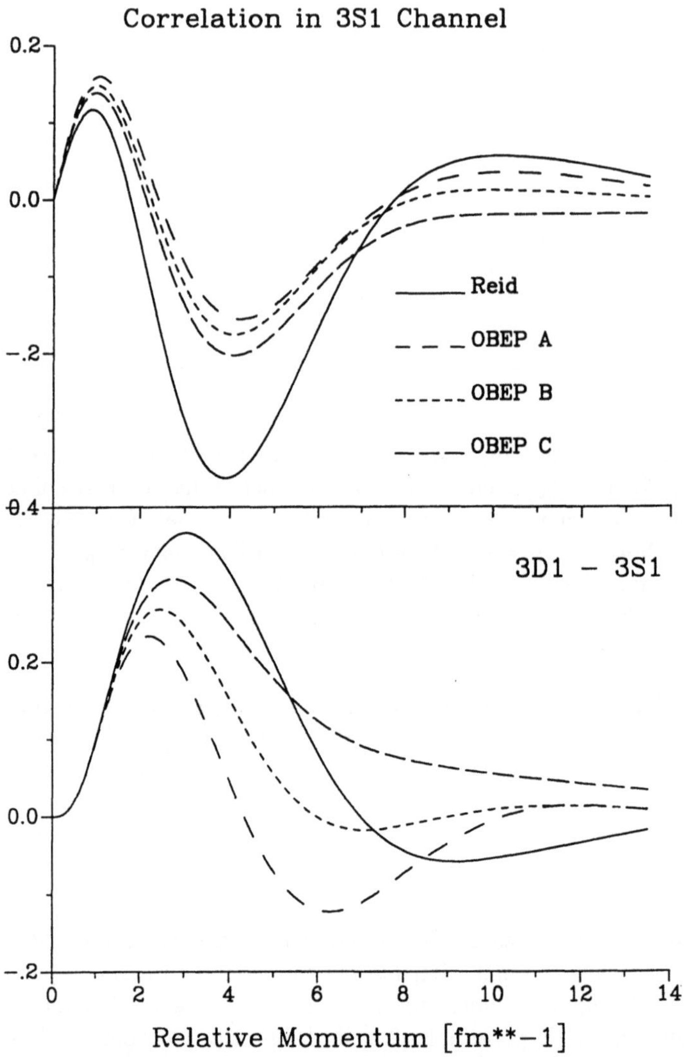

Fig. 2.5. The defect function in momentum space employing different NN interactions. The upper half presents results for the S-component of the $^3S_1$ correlation function [$l = l' = 0$ in (2.42)] and the lower half the corresponding D-component ($l = 2$, $l' = 0$). The abbreviations OBEP A, B, and C refer to the NN potentials A, B, and C defined in Table A.1 of [2.3].

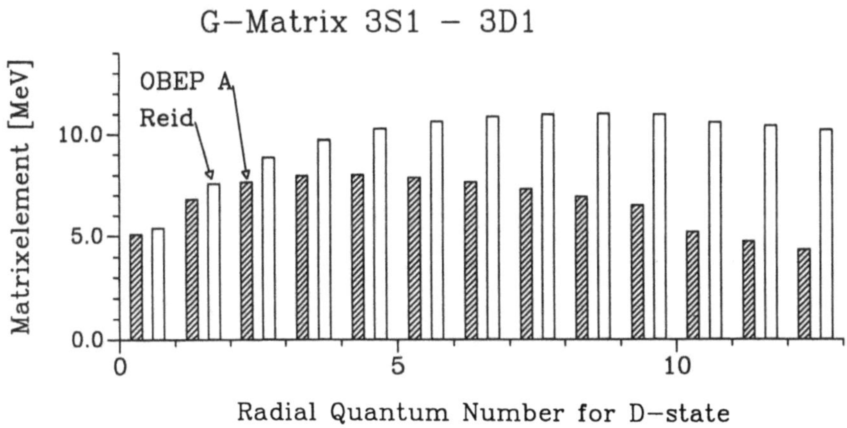

Fig. 2.6. Absolute values for the G-matrix elements $\langle n, {}^3D_1|G|n' = 0, {}^3S_1\rangle$ for different oscillator quantum numbers n. All matrix elements were calculated with the angle-averaged Pauli operator [see (2.34)] for ^{16}O, assuming an oscillator length b_{rel}=2.5 fm, a starting energy $W = -90$ MeV and $N_{\text{CM}} = L_{\text{CM}} = 0$.

for the coupled channels $^3S_1$–$^3D_1$ in Fig. 2.5. In the OBE model for the NN interaction, the tensor components are dominated by π exchange and (with opposite sign and shorter range) ρ exchange. The D-wave components for the defect functions seem to indicate that in the case of OBEA the short-range components of the π exchange have been reduced to such an extent that the tensor component originating from ρ exchange dominates the correlations at large relative momenta, leading to a different sign for the defect function.

Up to this point, we have discussed only some results related to the free-scattering matrix R. Such features like the reduction of the tensor components in the NN interaction can also be observed in the final G-matrix. As a first example, we discuss the matrix elements $\langle n, {}^3D_1|G|n' = 0, {}^3S_1\rangle$, which are visualized in Fig. 2.6. For the Reid potential (with a strong tensor component), these matrix elements are large even for large radial quantum numbers n. One consequence of this strong tensor component has been observed in [2.29]: The so-called core-polarization diagram displayed in Fig. 2.2(a) receives significant contributions to the effective interaction of nucleons from high-lying intermediate particle-hole states. The corresponding matrix elements calculated for OBEA decrease in absolute value with increasing n. Therefore, one may expect that the evaluation of the core-polarization diagram employing OBEA may converge better with respect to the energy of the intermediate particle-hole state than has been observed for the Reid potential [2.30].

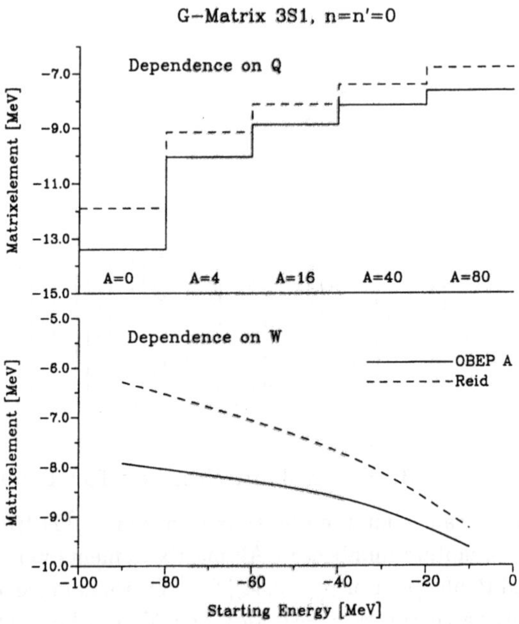

Fig. 2.7. Value for the G-matrix element $\langle n = 0,{}^3S_1|G|n' = 0,{}^3S_1\rangle$ calculated for various Pauli operators Q identified by the number of nucleons filling the lowest states in the oscillator model (*upper half*, $W = -30$ MeV) and for various starting energies W (*lower half*, Q for $A = 16$). The Reid potential (*dashed lines*) and the OBEA potential of [2.3] have been used. All matrix elements are calculated for an oscillator length $b_{\mathrm{rel}} = 2.5\ \mathrm{fm}^{-1}$ and with $N_{\mathrm{CM}} = L_{\mathrm{CM}} = 0$.

Another significant difference between the NN potentials under consideration is displayed in Fig. 2.7, where the diagonal matrix element $\langle n = 0,{}^3S_1|G|n' = 0,{}^3S_1\rangle$ is plotted for different Pauli operators (upper half) and as a function of the starting energy W (lower half). As has been discussed already in the introduction, the matrix elements are less attractive for a more restrictive Pauli operator (Pauli quenching) and a more attractive starting energy (dispersive quenching). As examples, we compare results obtained for the Reid and the OBEA potential. Ignoring the Pauli effects, the G-matrix should become identical for plane-wave states and the corresponding positive starting energy, since both interactions are fitted to NN phase shifts, that is, yield an identical scattering matrix. Although the matrix elements displayed in the lower half of the figure account for Pauli effects and are calculated for oscillator states, the two curves tend to merge for positive starting energies.

The dependence of the G-matrix on the starting energy is much weaker for the OBEA potential than for Reid. This can be related to the fact that the OBEA potential is softer with respect to the short-range components and

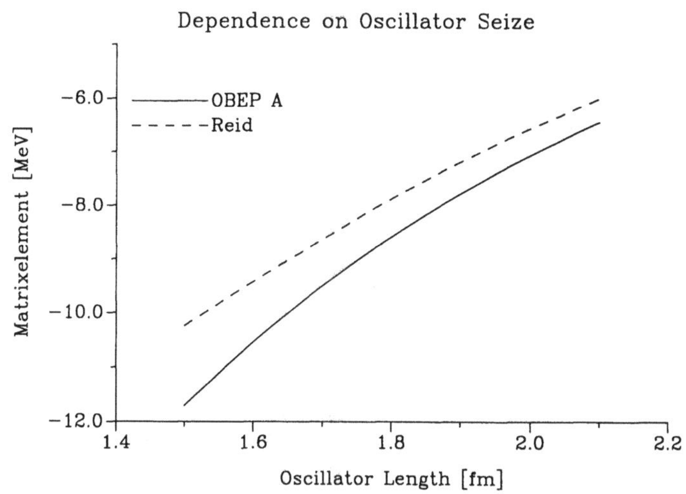

Fig. 2.8. Value for the G-matrix element $\langle n = 0, {}^3S_1|G|n' = 0, {}^3S_1\rangle$ as a function of the oscillator length $b = b_{\rm rel}/\sqrt{2}$. The Pauli operator is the angle-averaged Pauli of (2.34) appropriate for ^{16}O, and the starting energy is $W = -30$ MeV. For further details see Fig. 2.7.

contains a weaker tensor force. Therefore, the attractive components in G, which are due to terms in the Bethe–Goldstone equation (2.4) of second and higher order in V, are more important for the Reid potential than for OBEA. To yield the same phase shifts, the matrix elements of the bare potential V must be more attractive for OBEA. The dispersive effect only reduces the attractive components of second and higher order in V, and therefore it is larger for the Reid potential. Consequently, the G-matrix elements for negative starting energies are more attractive for potentials that are soft and contain a weak tensor force than for those that contain stronger short-range and tensor correlations. This effect explains the differences obtained in BHF calculations for the binding energies of nuclear matter [2.31] and finite nuclei [2.32] employing different realistic NN interactions.

While the dispersive quenching is quite different for the two NN interactions, the Pauli quenching seems to be quite similar, as can be seen from the upper half of Fig. 2.7. This may indicate that the small momentum components of the correlation, which are suppressed by the Pauli quenching, are similar for the different NN interactions (see also Fig. 2.5).

The dependence of the calculated matrix element on the oscillator length is also rather similar for OBEA and the Reid potential (see Fig. 2.8). The matrix elements are less attractive for a larger value of the oscillator length, which reflects the short range of the NN interaction.

Table 2.2. Some oscillator G-matrix elements $\langle nlSJ|G|n'l'SJ\rangle$ calculated for the Reid soft-core potential. All matrix elements have been calculated with an oscillator length b_{rel}=2.5 fm (b=1.7677), a starting energy $W = -50$ MeV, and an angle-averaged Pauli operator [(2.34)] for ^{16}O. The quantum numbers for the center-of-mass motion are $N_{CM} = L_{CM} = 0$.

l	l'	S	J	$n = n' = 0$	$n = 0, n' = 1$	$n = n' = 1$
0	0	1	1	−7.376	−6.350	−5.495
0	2	1	1	−5.340	−7.493	−5.268
2	2	1	1	1.225	1.101	1.367
1	1	0	1	1.291	1.963	3.060
0	0	0	0	−6.024	−4.915	−4.285
1	1	1	0	−1.814	−1.304	−1.182
1	1	1	1	1.862	2.112	2.810

Finally, in order to facilitate a test of the implementation of the program `CALGM`, we present some values for the G-matrix explicitly in Table 2.2.

2.6 Technical Note

The program `CALGM` is complete for the evaluation of the G-matrix employing the Reid soft-core potential [2.2]. In order to calculate the G-matrix for an OBE potential, one has to activate the statement "CALL BONN" in the subroutine `GENPOB`, link the subroutines `BONN` and those called by `BONN` to `CALGM`, and execute the resulting program setting the input parameter IINTER to 1 (see Table 2.3). `CALGM` expects input from the standard I/O unit 5. This input is described in Table 2.3. Furthermore, one may have to assign a disk file for temporary storage (Unit 8, see OPEN statement in the main program). The execution time for `CALGM` could be reduced by keeping the data, which for the present version are stored on the disk file 8, in RAM.

Note that the spline routine `SPLS3` uses the standard FORTRAN trick of declaring various arguments with fictitious dimensions of 1. For the arrays X and Y, this was changed to 2 to avoid some compilers complaining about the use of X(2) and Y(2).

A sample input file and the corresponding output are provided in `CALGM.INP` and `CALGM.OUT`.

Table 2.3: Input for the Program `CALGM`. All lines use the same format: `7(5X,I2)/2(5X,F10.5)` Note that "line 3" in reality consists of one line of seven integers, followed by two floating-point numbers on the next line. As the input is sufficiently complicated to keep it in a file, using a format allows one to devote the latter part of each line to comments.

Line	Variable	Explanation
1	IPRI(1)	Print option, G-matrix elements are printed for radial quantum numbers in the range $0 \leq n, n' \leq$IPRI(1) $- 1$
	IPRI(2)	Print option, the matrix elements of the Pauli operator are printed if IPRI(2) > 0
	IPRI(3)	Print option, the mesh points in momentum space are printed if IPRI(3) > 0
	IPRI(4)	Calculate and print the defect function (2.42) for oscillator states up to n' = IPRI(5) $- 1$
	IINTER	If IINTER = 1, employ the Reid soft-core potential, else OBEP
2	IWIED	IWIED=0 set up mesh points, calculate V and G IWIED=1 calculate V and G (mesh points as before) IWIED=2 calculate G (V and mesh points as before) IWIED=3 stop
3	NMAX	Maximal radial oscillator quantum number, to be considered in (2.36) [NMAX > IPRI(1), typical value: NMAX=20]
	LORB(1)	Orbital angular momentum for the relative motion in the partial wave under consideration
	LORB(2)	= LORB(1) for uncoupled channels, =LORB(1)+2 for coupled channels
	ISPIN	= Spin of the interacting nucleons
	JTOTAL	= Total angular momentum for the partial wave
	NCM	$N_{\mathrm{CM}} + 1$
	LCM	L_{CM}
	OSCLSP	Oscillator length $b = b_{\mathrm{rel}} / \sqrt{2}$ in fm
	OMEGEN	Starting energy W in MeV
4	IPAULI	=0: No Pauli Operator, =1: Eden–Emery approximation (2.35), =2: angle-averaged approximation (2.34)
	NEDEN	Corresponds to N_{EE} in (2.35)
	NPAMA1	For Pauli operator in angle-averaged approximation: Q=1 if ($2 * n_1 + l_1 >$ NPAMA1 and $2 * n_2 + l_2 >$ NPAMA1) and ($2 * n_1 + l_1 >$ NPAMA2 or $2 * n_2 + l_2 >$ NPAMA2)
	NPAMA2	See above
5	IWIED	Continue calculation with more input lines 3–5 or stop; see definition of IWIED in line 2

References

[2.1] T. Hamada and I. Johnston, Nucl. Phys. **34**, 382 (1962)
[2.2] R.V. Reid, Ann. Phys. (N.Y.) **50**, 411 (1968)
[2.3] R. Machleidt, Adv. in Nucl. Phys. **19**, 189 (1989)
[2.4] R. Vinh Mau, in *Mesons in Nuclei* Vol. I (North-Holland, Amsterdam, 1979) p. 151
[2.5] H. Müther, Prog. in Part. and Nucl. Phys. **17**, 97 (1986)
[2.6] B.D. Day, Phys. Rev. C**24**, 1203 (1981)
[2.7] B.D. Day and R.B. Wiringa, Phys. Rev. C**32**, 1057 (1985)
[2.8] P.J. Brussard and P.W.M. Glaudemans, *Shell Model Applications in Nuclear Spectroscopy* (North-Holland, Amsterdam, 1977)
[2.9] H. Feshbach, Ann. of Phys. **19**, 287 (1962)
[2.10] C. Bloch and H. Horowitz, Nucl. Phys. **8**, 51 (1958)
[2.11] B.H. Brandow, Rev. Mod. Phys. **39**, 771 (1967)
[2.12] H. Müther, A. Polls, and T.T.S. Kuo, Nucl. Phys. A**435**, 548 (1985)
[2.13] T.T.S. Kuo and G.E. Brown, Nucl. Phys. **85**, 40 (1966)
[2.14] T.T.S. Kuo and E. Osnes, *Folded Diagram Theory of the Effective Interaction in Atomic Nuclei*, Lecture Notes in Physics 364 (Springer-Verlag, Berlin and Heidelberg, 1990)
[2.15] H.A. Bethe, B.H. Brandow and A.G. Petschek, Phys. Rev. **129**, 225 (1962)
[2.16] J.D. Jeukenne, A. Lejeune, and C. Mahaux, Phys. Rep. C**25**, 83 (1976)
[2.17] T.T.S. Kuo, Z.Y. Ma, and R. Vinh Mau, Phys. Rev. C**33**, 717 (1987)
[2.18] J.G. Zabolitzky, Phys. Lett. B**47**, 487 (1973)
[2.19] P.U. Sauer, Nucl. Phys. A**150**, 467 (1970)
[2.20] M.I. Haftel and F. Tabakin, Nucl. Phys. A**158**, 1 (1970)
[2.21] R.K. Tripathi, A. Faessler, and H. Müther, Phys. Rev. C**10**, 2080 (1974)
[2.22] C.W. Wong, Nucl. Phys. A**91**, 399 (1967)
[2.23] M. Moshinsky, Nucl. Phys. **13**, 104 (1959)
[2.24] R.J. Eden and V.J. Emery, Proc. Roy. Soc. A**248**, 266 (1958)
[2.25] M. Abramowitz and I.A. Stegun, *Handbook of Mathematical Functions* (Dover, New York, 1964)
[2.26] B.R. Barret, R.G.L. Hewitt, and R.J. McCarthy, Phys. Rev. C**3**, 1137 (1971)
[2.27] E.W. Krenciglowa, C.L. Kung, T.T.S. Kuo, and E. Osnes, Ann. Phys. (N.Y.) **101**, 154 (1976)
[2.28] R.L. Becker, in *Effective Interactions and Operators in Nuclei*, edited by B.R. Barret (Springer-Verlag, Berlin and Heidelberg, 1975) p. 96
[2.29] J.P. Vary, P.U. Sauer, and C.W. Wong, Phys. Rev. C**7**, 1776 (1973)
[2.30] H.M. Sommermann, H. Müther, K.C. Tam, T.T.S. Kuo, and A. Faessler, Phys. Rev. C**23**, 1765 (1981)
[2.31] F. Coester, S. Cohen, B.D. Day, and C.M. Vincent, Phys. Rev. C**1**, 769 (1970)
[2.32] R. Machleidt, H. Müther, and A. Faessler, Nucl. Phys. A**241**, 18 (1975)

3. The Nuclear-Matter Effective Interaction

Martin Fuchs and Philip J. Siemens

3.1 Background

The independent pair approximation is the simplest *ab initio* theory that can describe the normal ground state of a Fermi liquid bound by local short-range interactions. It is the second step in a hierarchy that begins with independent particle motion in a mean field, and may be refined to include the simultaneous correlations of three, four, or more particles. The simpler mean-field approximation does not allow a thermodynamic limit of large bound systems of uniform density, because the average interaction has to be either repulsive, precluding binding, or attractive, producing collapse. Pairwise correlations allow the motion of the fermions to avoid the short-range repulsive forces that stabilize the matter against collapse while winning binding energy from attractive forces at somewhat larger distances. While the treatment of more complicated correlations is necessary to obtain a precise description of the ground state, two-body correlations are enough to give qualitatively correct and, in the case of nuclear matter, quantitatively reasonable results, which may then be corrected if necessary for higher-order correlations.

Pair correlations may be implemented either in multiple-scattering theory by medium-dependent effective interactions, or in variational theory, by a Jastrow-type wave function. The former approach, historically the first to be applied quantitatively, has the advantage that the effective interactions may be used conveniently as the starting point for many other applications, such as theories of collective motion. In addition, the scattering-theory equations for the effective interactions have a linear structure that makes them much simpler to solve than the non-linear hypernetted-chain equations needed to get useful results from variational wave functions. Readable presentations of the Brueckner multiple-scattering theory of pair correlations can be found in [3.1–5]. The code presented here is an implementation following closely the method of Legindgaard [3.6]. It requires as input an off-shell scattering matrix, which may be obtained, for the case of a local potential, by an included auxiliary program that uses methods introduced by Bethe, Brandow, and Petschek [3.7].

The scattering in free space of two particles interacting via a two-body interaction V is described by the irreducible four-point vertex operator T_E, which satisfies the Lippmann–Schwinger integral equation

$$T_E = V + VG_ET_E, \tag{3.1}$$

where G_E is the free two-body propagator for energy E. T_E has the desirable property that its matrix element between plane waves gives the exact scattering amplitude, while the corresponding matrix element of V would only give the first Born approximation to the scattering. Brueckner [3.8] proposed to describe the scattering of two particles in an interacting system by modifying the propagator in (3.1) to include the effects of the surrounding medium. The changes in the propagator may be understood in terms of mean-field theory (see [3.3,5,8,9]) and include two effects:

- the potential energy of the particles due to their interaction with the medium,
- the exclusion of intermediate states already occupied by other fermions.

The resulting operator T_E may be regarded as an effective interaction whose matrix elements between independent-particle or mean-field states include the correlations of the interacting pair of fermions to all orders in V. This result is most readily understood in terms of the Møller wave operator $\Omega_E \equiv (1 - G_E V)^{-1}$, since

$$T_E = V\Omega_E, \tag{3.2}$$

while the projection of the many-body wave function onto the two particles on which T_E operates is [3.4]

$$|\psi\rangle = \Omega_E|\Phi\rangle, \tag{3.3}$$

where $|\Phi\rangle$ is the corresponding projection of the independent-particle Slater determinant.

In principle, the independent-particle propagator G_E should be chosen to correspond to the system being studied, for example a nucleus of N neutrons and Z protons in its ground state. This function of four spatial variables could be readily constructed from products of the two-point propagators of the independent-particle model. The resulting T_E, however, would be a function of four three-dimensional coordinates or momenta without the simplification of factorization. Though of course T_E possesses some discrete symmetries related to interchanging identical particles, to time reversal, and possibly to parity and rotation if the corresponding nuclear state has these symmetries, its representation in a computer would involve a very large matrix. Instead, T_E is often approximated by a density-dependent interaction, appealing to the argument that the correlations are rather short range and therefore could be expected to depend mainly on the properties of the matter in the immediate vicinity of the interacting particles. The density-dependent interaction can then be taken from computations in uniform matter, which are greatly expedited by translational symmetry. The present code performs this computation in uniform matter.

3.2 The Bethe–Goldstone Equation in Uniform Matter

Translational symmetry simplifies computations of T_E in two important ways:

- The independent-particle wave functions are plane waves $|\boldsymbol{p}\rangle$, eigenstates of momentum with eigenvalue $\boldsymbol{p}$. Having known simple wave functions is a significant convenience.
- The average momentum $\boldsymbol{K} \equiv \frac{1}{2}(\boldsymbol{p}_1 + \boldsymbol{p}_2)$ of the interacting pair is conserved. As a result, the scattering equation can be solved within a subspace of fixed total momentum, which reduces the dimensionality of the integrals or sums over basis states to a single three-dimensional variable representing the relative momentum $\boldsymbol{k} \equiv \frac{1}{2}(\boldsymbol{p}_1 - \boldsymbol{p}_2)$. This is an essential advantage that makes the case of uniform matter much more tractable than a finite system.

As an added convenience, the isotropy of infinite matter implies that two-body operators can only depend on the magnitudes of $\boldsymbol{K}$ and the initial and final relative momenta $\boldsymbol{k}$ and $\boldsymbol{k}'$, and on the angles between them, and not on the directions of $\boldsymbol{K}$, $\boldsymbol{k}$, and $\boldsymbol{k}'$ separately. The most natural basis for the two-particle operators and wave functions is thus the antisymmetrized products of plane waves,

$$|\boldsymbol{p}_1, \boldsymbol{p}_2\rangle = |\boldsymbol{k} + \boldsymbol{K}, \boldsymbol{k} - \boldsymbol{K}\rangle = |\boldsymbol{K}\ \boldsymbol{k}\rangle = |(\boldsymbol{p}_1 + \boldsymbol{p}_2)(\boldsymbol{p}_1 - \boldsymbol{p}_2)\rangle. \tag{3.4}$$

In this basis, the uniform-matter two-particle propagator is

$$\langle \boldsymbol{K}\boldsymbol{k}'|G_E|\boldsymbol{K}\boldsymbol{k}\rangle = \delta(\boldsymbol{k}' - \boldsymbol{k})\frac{[1 - n(\boldsymbol{k} - \boldsymbol{K})][1 - n(\boldsymbol{k} + \boldsymbol{K})]}{E - E(\boldsymbol{k} - \boldsymbol{K}) - E(\boldsymbol{k} + \boldsymbol{K})}, \tag{3.5}$$

where $n(\boldsymbol{p})$ is the single-particle occupation number for momentum $\boldsymbol{p}$. In the ground state, $n(\boldsymbol{p}) = \Theta(|\boldsymbol{p}| - p_F)$, where p_F is the Fermi momentum, and the Heaviside function Θ is one for positive values of its argument and zero for negative argument.

The independent-particle Hamiltonian H_0 is characterized by the function $E(p)$, which gives the single-particle energy as a function of the momentum p. The choice of this function has long been controversial. Bethe and Goldstone [3.10] recognized that a self-consistent or Hartree–Fock choice of H_0 guarantees the cancellation of the simplest self-energy insertions for occupied (hole) states but not for unoccupied (particle) states, and argued that H_0 could conveniently be chosen for unoccupied states as the pure kinetic energy without potential contributions. Much later, Bethe [3.2] argued that this choice would make the contributions of three-body correlations small for realistic nuclear forces. Later computations [3.11] showed that the sum of three- and four-body correlation effects, while smaller than the effect of two-body correlations, give substantial contributions to the binding of nuclear matter. Meanwhile, Mahaux and collaborators [3.3] argued that a self-consistent choice of intermediate-state energies not only gives a somewhat better estimate of the binding energy, but more importantly gives a more reasonable account of the correlations involving momenta near the Fermi level, which are suppressed by the big energy gap of Bethe's prescription. The present code treats $E(p)$ as input, to be chosen by the user.

To take advantage of the isotropy of nuclear matter, a partial-wave decomposition is introduced:

$$\langle \boldsymbol{K}klSJM|\boldsymbol{K}\boldsymbol{k}\rangle = \sum_{mm'} \langle lSmm'|JM\rangle Y_l^m(\Omega_k)\chi_S^{m'}, \tag{3.6}$$

where Y_l^m and $\chi_S^{m'}$ are, respectively, eigenfunctions of orbital and spin angular momentum, quantized along the average momentum $\boldsymbol{K}$. The potential V, of course, is rotationally invariant and therefore diagonal in $\boldsymbol{K}$, J, S, and M and independent of M; the nuclear force has a strong tensor component that connects states with $l = J \pm 1$. The propagator G_E of (3.5) is diagonal in $\boldsymbol{K}$, S, and M. In the customary and quite accurate [3.6] angle-averaged approximation for the propagator, the operators are also diagonal in J and are independent of M.

The most important effect of the correlations in T_E is to suppress the effect of short-range repulsion in V corresponding to a reduction in the pair wave function at short distances. This effect is already present in free-space scattering, where it is sufficient to produce a net attractive effective interaction that could bind nuclear matter. The main effect of the medium on the short-range correlations is the "spectral correction" due to the modification of H_0 in the medium. Since the short-range correlations involve Fourier components of high momentum, they are rather insensitive to the exclusion of low-momentum occupied states, or to the details of the spectrum $E(p)$ in the medium. These affect mainly the intermediate-range correlations, which in free space build up the wave function at distances where the force is attractive. These intermediate-range correlations are inhibited in matter by the exclusion of Fourier components less than the Fermi momentum p_F, as well as by the fact that in nuclear matter more energy is needed to increase a particle's momentum than is necessary in free space. Because these effects increase with the density of the medium, they are responsible (together with the exchange nature of the nuclear force) for the decrease of nuclear attraction with increasing density and therefore provide an important contribution to the saturation of nuclear binding at the observed density of heavy nuclei. Multiparticle correlations and multibody forces also have major influences on the saturation density; these effects are beyond the scope of the present method, but can be minimized or mimicked by an appropriate choice of H_0 [3.3,11].

3.3 The Reference Spectrum Method

The need to describe short-range correlations places stringent demands on the method used to solve the scattering equation: it has to include high-momentum components with great precision in order to fit the correlations to the interactions. Since these correlations are mostly influenced by the single-particle potential but not much affected by the antisymmetrization, it is convenient to deal with these effects by solving the equation in two stages. In the first stage, an off-shell or "reference" interaction T_E^R is defined by the equation [3.7],

$$T_E^R = V + VG_E^R T_E^R, \tag{3.7}$$

where the "reference" Green's function G_E^R is simply a free-particle propagator with a shifted energy and mass,

$$G_E^R = (E - H_R)^{-1}, \tag{3.8}$$

with H_R given by the simplified single-particle "reference" spectrum

$$E_R(p) = \frac{\boldsymbol{p}^2}{2m^*} + A. \tag{3.9}$$

Once T_E^R is known, T_E can then be found from the identity [3.7]

$$T_E = T_E^R + T_E^R(G_E - G_E^R)T_E, \tag{3.10}$$

which is more convenient to solve than (3.1) because T_E^R already has the short-range correlations built in and therefore has much milder short-range repulsion than V.

The reference interaction may be solved by the same techniques used to solve the free-scattering problem. If V is not local, the method must be supplied by the user, but can probably be readily adapted from whatever method was used in fitting V to the measured phase shifts. For a local potential, the coordinate representation may be convenient, and a program to perform this calculation is included. The basis transformation to radial coordinates is

$$\langle \boldsymbol{K} rlSJM | \boldsymbol{K} kl'S'J'M' \rangle = \delta_{ll'}\delta_{SS'}\delta_{JJ'}\delta_{MM'} i^l \sqrt{\frac{2}{\pi}} \frac{J_l(kr)}{kr}, \tag{3.11}$$

where $J_l(x) \equiv x j_l(x)$ is a Bessel function. The radial wave functions are then given by

$$u_{ll'JS}^R(k,r) = i^l \sqrt{\frac{2}{\pi}} kr \langle \boldsymbol{K} rlSJM | \Omega_E^R | \boldsymbol{K} kl'SJM \rangle, \tag{3.12}$$

where Ω_E^R is the wave operator corresponding to T_E^R. They satisfy the radial wave equation [3.7],

$$\begin{aligned}\left[\frac{\mathrm{d}^2}{\mathrm{d}r^2} - \frac{l(l+1)}{r^2} - \gamma^2\right]\left[\delta_{ll'} J_l(kr) - u_{ll'JS}^R(k,r)\right] = \\ -\sum_{l''} m^* V_{ll''JS}(r) u_{l''l'JS}^R(k,r),\end{aligned} \tag{3.13}$$

where

$$\gamma^2 \equiv K^2 + (2A - E)m^*/\hbar^2, \tag{3.14}$$

and $V_{ll'JS}(r)$ is the representation of the potential,

$$\langle \boldsymbol{K} rlSJM|V|\boldsymbol{K} r'l'SJM\rangle \equiv V_{ll'JS}(r)\delta(r-r'). \tag{3.15}$$

The radial wave functions are regular at $r = 0$ and satisfy the boundary condition $u^R_{ll'JS}(k,r) \rightarrow \delta_{ll'}J_l(kr)$ as $r \rightarrow \infty$.

Once the radial wave equation has been solved, the matrix elements of the reference interaction are obtained by implementing (3.2) in the partial-wave basis,

$$\begin{aligned} T^R_{ll'JS}(k,k') \equiv & \langle \boldsymbol{K} klSJM|T^R_E|\boldsymbol{K} k'l'SJM\rangle \\ = & \frac{4\pi}{kk'}\sum_{l''}\int \mathrm{d}r\, J_l(kr)V_{ll''JS}(r)u^R_{l''l'JS}(k',r). \end{aligned} \tag{3.16}$$

The resulting matrix is real and symmetric, since T^R_E is Hermitian. In the case when the potential includes a hard core of radius r_c, the product $V(r)u(r)$ is indeterminate for $r \leq r_c$. The ambiguity may be resolved by first solving the hard-core potential alone and then including the outer potential via the two-potential form of the Lippmann–Schwinger equation [3.7]; the result is that the radial integrals in (3.16) are given by

$$\begin{aligned} &\langle \boldsymbol{K} klSJM|T^R_E|\boldsymbol{K} k'l'SJM\rangle \\ &= \frac{4\pi}{kk'}\Bigg(\sum_{l''}\int_{r_c}^{\infty} \mathrm{d}r[J_l(kr) - H_l(k,\gamma r)]V_{ll''JS}(r)u^R_{l''l'JS}(k',r) \\ &\quad + \delta_{ll'}\Big\{(\gamma^2 + k'^2)\int_0^{r_c} \mathrm{d}r\, J_l(kr)J_l(k'r) \\ &\qquad + J_l(kr_c)\frac{\partial}{\partial r}[J_l(k'r) - H_l(k',\gamma r)]_{r_c}\Big\}\Bigg), \end{aligned} \tag{3.17}$$

where H_l is the decaying spherical Hankel function of order l, normalized to J_l at r_c:

$$H_l(k,\gamma r) \equiv H_l^{(-)}(\gamma r)\frac{J_l(kr_c)}{H_l^{(-)}(\gamma r_c)}, \tag{3.18}$$

with $H_l^{(-)}(x) = \mathrm{i}^{l+1}(\mathrm{i}x)h_l^{(1)}(\mathrm{i}x)$. The corrections in (3.17) reflect the non-locality of the off-shell equation for T^R_E .

3.4 The Legindgaard Representation

When Bethe, Brandow, and Petschek originally introduced the reference spectrum [3.7], they used it to approximate H_0 for all unoccupied states. For the modern self-consistent choices of single-particle spectra, this approximation is not very good because the shift A and effective mass m^* needed to represent the attractive spectrum just above the Fermi momentum imply a strong repulsive single-particle potential for large momenta. By contrast, both self-consistent calculations and experimental observations of scattering of intermediate-energy nucleons from nuclei indicate only a mild repulsion, which may even fall off for high momenta. However, the reference spectrum can probably give a reasonable representation of H_0 for momenta above about 3 fm^{-1}, where the potential energy is anyway quite small compared with the kinetic energy. Legindgaard [3.6] observed that, if H_R is chosen to match H_0 above some momentum $p_{\max} > p_F$, then the difference of the propagators in (3.9) vanishes outside a finite subspace. As a result, (3.9) need only be solved within a finite subspace that includes the momenta for which the difference of the propagators $G_E^R - G_E$ is non-vanishing. The advantages of restricting the problem to a finite subspace are remarkable:

- the compactness of the subspace permits an expansion in a discrete set of orthogonal functions, and
- the smooth dependence of the effective interactions on this limited range of momenta allows them to be accurately represented by a small number of expansion coefficients.

A convenient choice of this subspace is, for example, a sphere in the relative momentum of radius $k_{\max}$ as long as $k_{\max} > K + p_{\max}$. Then the orthogonal functions can be chosen as polynomials in the relative momenta. It is convenient to include in the basis the usual entrance-channel decomposition of the relative angular momentum (3.6). Quantizing the angular momenta along the average momentum $\boldsymbol{K}$ takes advantage of the common symmetries of the two-body interaction and G_E, so that all operators are diagonal in $\boldsymbol{K}$, S, M, and parity $(-1)^l$. The relative radial momentum may be represented by orthonormal polynomials $F_n(k)$,

$$\langle \boldsymbol{K}klSJM|\boldsymbol{K}nlSJM\rangle = F_n(k). \tag{3.19}$$

The F_n satisfy orthonormality and completeness conditions,

$$\int_0^{k_{\max}} k^2 \mathrm{d}k\, F_n(k)F_{n'}(k) = \delta_{nn'} \tag{3.20}$$

and

$$\sum_{n=1}^{\infty} F_n(k)F_n(k') = \delta(k'-k)/k^2, \tag{3.21}$$

and are polynomials of order $n-1$ given by

$$F_n(k) = \frac{\sqrt{2n+1}}{k_{\max}^{3/2}}(-1)^{n-1}\sum_{s=0}^{n-1}\frac{(-1)^s(n+s+1)!}{s!(n-1-s)!(s+2)!}\left(\frac{k}{k_{\max}}\right)^2. \tag{3.22}$$

These polynomials are chosen to give a rapidly converging representation of T_E, retaining only terms with $n \leq n_{\max}$. However, neither they nor any other smooth function will be able to give an accurate description of the propagators, which are diagonal in relative momentum k and therefore require many polynomials to obtain an accurate representation of the δ-function (3.21). Furthermore, even considered as a function of k, G_E is not analytic at the Fermi surface. Luckily, the rapid convergence of the expansion of T_E with $n_{\max}$ is sufficient to give convergence of the solutions of (3.10) with $n_{\max}$, as shown in [3.6].

In this basis representation, (3.10) becomes

$$\begin{aligned}
&\langle \boldsymbol{K} n_1 l_1 S J_1 M | T_E | \boldsymbol{K} n_2 l_2 S J_2 M\rangle = \langle \boldsymbol{K} n_1 l_1 S J_1 M | T_E^R | \boldsymbol{K} n_2 l_2 S J_2 M\rangle \\
&\quad + \sum_{\substack{n_3 l_3 J_3 \\ n_4 l_4 J_4}} \langle \boldsymbol{K} n_1 l_1 S J_1 M | T_E^R | \boldsymbol{K} n_3 l_3 S J_3 M\rangle \\
&\quad \times \langle \boldsymbol{K} n_3 l_3 S J_3 M | (G_E - G_E^R) | \boldsymbol{K} n_4 l_4 S J_4 M\rangle \\
&\quad \times \langle \boldsymbol{K} n_4 l_4 S J_4 M | T_E | \boldsymbol{K} n_2 l_2 S J_2 M\rangle. \qquad (3.23)
\end{aligned}$$

The angular-momentum sums are not extensive: for $S = 0$, $l = J$; while for $S = 1$, basis states with $l = J \pm 1$ are coupled by (3.23). In the customary and quite accurate angle-averaged approximation for the propagator, the operators are also diagonal in J and are independent of M. Moreover, the matrices are symmetric because the operators are Hermitian. Thus, the scattering equation is realized, for each choice of K, S, M, and parity, by a matrix equation of dimension $N_\mathrm{D} \times N_\mathrm{D}$, where $N_\mathrm{D} = n_{\max} J_{\max}$ for $J_{\max}$, the maximum number of angular-momentum couplings considered (about half the maximum angular momentum because of parity restrictions). In the angle-averaged approximation, $J_{\max} = 1$.

The scattering equation (3.10) may be formally solved in the form

$$T_E - T_E^R = \left\{ \left[1 - T_E^R (G_E - G_E^R) \right]^{-1} - 1 \right\} T_E^R. \qquad (3.24)$$

Thus, it is merely necessary to invert an $N_\mathrm{D} \times N_\mathrm{D}$ matrix once the reference matrix T_E^R has been computed. The matrix inversion is not demanding, since the matrix representing the operator in square brackets in (3.24) is symmetric and diagonally dominated.

The numerical work then consists of four steps:

- obtain the matrix elements of T_E^R in the basis of relative momentum and angular momentum;
- transform T_E^R and $G_E - G_E^R$ into the Legindgaard polynomial representation;
- realize the right-hand side of (3.24) by matrix multiplication and inversion;
- transform $T_E - T_E^R$ back into the momentum representation, then add T_E^R to get T_E.

These steps are implemented in the accompanying computer codes.

3.5 Numerical Methods

Two computer codes are supplied. The first computes the off-shell scattering amplitude T_E^R for a local potential. The second solves (3.10) by transforming to the Legindgaard basis and back.

3.5.1 The Off-Shell Reference Effective Interaction for a Local Potential

The code TREF solves the off-shell reference scattering equation (3.7) in coordinate representation by computing the radial wave functions of the relative motion for a two-body potential that is local in the radial variable r, but may have an arbitrary dependence on angular momentum subject only to the conservation of total angular momentum, isospin (or equivalently spin), and parity. Subroutines are included for the Reid soft-core potential [3.12], which may readily be replaced by another potential, given by its projections into each angular-momentum channel. The results are given as momentum-space matrix elements of the effective interaction T_E^R for values of the momentum specified by the user.

The radial wave functions are found by solving the boundary-value problem on a discretization mesh, possibly with variable step size, using a Numerov integration combined with a Newton method to find the suitable initial conditions. The boundary condition at small radius is enforced by integrating the inhomogeneous equation outwards to a large radius where it is matched with the asymptotic analytical solution (Bessel function). Where the step size changes, the connection is made with a sixth-order polynomial extrapolation. To ameliorate the strong radial singularity of the tensor operator, an artificial hard core is necessary. To safely suppress any contribution from the homogeneous solution, it is also retained for the single channels. The matrix elements are then obtained using the separation treatment of [3.7]. The inhomogeneous term in the differential equations and the wave function outside the core in the final state involve spherical Bessel and Hankel functions, which are evaluated using the analytical expressions in sin, cos, and exp, or by recursion schemes.

Once the radial wave functions have been obtained, their matrix elements (3.16–17) are evaluated with trapezoidal integration using the same radial mesh as for the differential equation. These are then output to separate files for each angular-momentum channel.

3.5.2 The Nuclear Matter Reaction Matrix

The code TNM solves the scattering equation (3.10) for uniform nuclear matter. Input and output are given in momentum–angular momentum representation, but the computation is performed in the Legindgaard polynomial basis. The angle-averaged propagator is used, but the computations are grouped so that the code could readily be expanded to include the angular couplings generated by the full propagator. Inputs are the reference interaction $T^R_{ll'JS}(k,k')$ and the parameters describing it, and the density and single-particle spectrum of the nuclear matter along with the desired center-of-mass momentum and starting energy E. These inputs have to be given over a sufficient range of momenta to include the desired subspace of the Legindgaard method. It should be remarked that the input reference interaction ought to be computed with fixed starting energy E.

The transformation of the input $T^R_{ll'JS}(k,k')$ to the Legindgaard basis is performed by integration on the momentum mesh of the input:

$$\begin{aligned}&\langle \boldsymbol{K}nlSJM|T^R_E|\boldsymbol{K}n'l'SJM\rangle \\ &\quad = \mathrm{i}^{l-l'}\frac{4\pi}{(2\pi)^3}\int_0^{k_{\max}} k^2\mathrm{d}k \int_0^{k_{\max}} k'^2\mathrm{d}k' F_n(k)F_{n'}(k')T^R_{ll'JS}(k,k'). \qquad (3.25)\end{aligned}$$

The Legindgaard representation of the difference of the propagators (which is diagonal in momentum) is found similarly, with interpolation needed only for the single-particle energies of the nuclear-matter propagator. A ten-point Gauss–Legendre integration is used for the propagators. Matrix inversion is by LU decomposition with back-substitution [3.13]. The right-hand side of (3.24) is computed, and the resulting $T_E - T^R_E$ is then transformed back into the momentum representation by

$$\begin{aligned}&T_{ll'JS}(k,k') - T^R_{ll'JS}(k,k') \\ &\quad = \mathrm{i}^{l-l'}\frac{(2\pi)^3}{4\pi}\sum_{nn'}^{n_{\max}} F_n(k)F_{n'}(k')\langle \boldsymbol{K}nlSJM|T_E - T^R_E|\boldsymbol{K}n'l'SJM\rangle. \qquad (3.26)\end{aligned}$$

The final output T_E is then obtained by adding the input T^R_E.

3.6 Remarks on the Code

The code implements the procedures described above in a straightforward way. Most of the variables are passed via COMMON, which is divided in blocks; variables of physical interest or that would be changed from run to run are included in the subroutine call lists. The DIMENSION allocations all use parameter variables, which may be specified by the user. With the non-averaged Pauli operator in mind, the Legindgaard matrices in `TNM` are constructed to contain all the components of T^R_E with the same spin and parity quantum numbers; for their size, the user should consult the discussion in Sect. 3.3 above.

Before `TNM` can be run, files with the reference interaction must be prepared either by running `TREF` or, in the case of non-local free-space interactions, from other programs supplied by the user. The format of these files is specified in Sect. 3.5.3 below.

3.6.1 Input to TREF

For computations using the Reid soft core potential, only five inputs are needed:

- the value of m^*/m;
- the value of γ^2 [see (3.14)] in fm^{-2};
- a three-letter identifying prefix for the files to be passed to TNM;
- two variables NCHLO and NCHUP specifying the range of angular-momentum channels to be computed; the correspondence between internal channel numbers and the angular momenta $ll'JS$ is given in Table 3.1.

These variables are given in a BLOCK DATA unit at the beginning of the main program TREF, which is supplied with default values (1.0D0, 1.4D0, ''TRF'', 1, 12).

Table 3.1. Channel labels for TREF and TNM. NCH is the internal integer label, "File Name" indicates the output/input files, where "TRF" is the character string which may be changed in the BLOCK DATA statement (see text).

S	Π	$\langle JlS\|T\|J'l'S'\rangle$	NCH	File Name
0	1	$\langle 000\|T\|000\rangle$	1	"TRF"000
0	−1	$\langle 110\|T\|110\rangle$	2	"TRF"110
1	−1	$\langle 011\|T\|011\rangle$	3	"TRF"011
1	−1	$\langle 111\|T\|111\rangle$	4	"TRF"111
0	1	$\langle 220\|T\|220\rangle$	5	"TRF"220
1	1	$\langle 221\|T\|221\rangle$	6	"TRF"221
1	−1	$\langle 211\|T\|211\rangle$	7	"TRF"211
1	−1	$\langle 231\|T\|231\rangle$	8	"TRF"231
1	−1	$\langle 231\|T\|211\rangle$	9	"TRF"21C
1	1	$\langle 101\|T\|101\rangle$	10	"TRF"101
1	1	$\langle 121\|T\|121\rangle$	11	"TRF"121
1	1	$\langle 121\|T\|101\rangle$	12	"TRF"10C

For computations with a user-supplied local potential, four subroutines need to be replaced. DIRPOT1 and DIRPOT2 compute the potential in each angular-momentum channel in units fm^{-2}, which may be converted from MeV units using the conversion constant $m/\hbar^2 = 41.47$ MeV fm^2. The values are returned from FUNCTION entries according to the scheme $V_{lJS}(r) = \mathrm{DIRPOT1}(J, l, s, r)$ for uncoupled channels, $V_{ll'JS}(r) = \mathrm{DIRPOT2}(J, l, r)$ for $l = l'$, and $V_{ll'JS}(r) = \mathrm{TENPOT}(J, r)$ for $l \neq l'$ for tensor-coupled channels. It may then also be necessary to change the radial and momentum meshes defined in RGRID and KGRID, which puts the required values in common blocks RMESH and KMESH, along with integers MINR, MAXR, MINK, and MAXK specifying the beginning and ending points of the desired integration regions. RMESH also specifies the hard core to be in the position MHCO. Note that the value of γ^2 has to be chosen consistent with the starting energy, center-of-mass momentum, and single-particle spectrum subsequently used in TNM. Simpson's rule for the transformation to the polynomial representation in TNM requires that the input momentum mesh contains an *even* number of mesh points.

3.6.2 Output from TREF

Most of the output from TREF is contained in a series of files designed as input to TNM and is described in the next section. The only other output, to the screen, is a measure $\mathrm{SD}(l, l', J, S)$ of the accuracy of the solution of the differential equations. This measure is based on the observation that while T_E^R is Hermitian and therefore symmetric, the method used to compute T_E^R is not symmetric but distinguishes the "entrance channel," that is, the bra, from the adjoint ket. The difference between T_E^R and its transpose is thus an indication of the quality of the solutions. If the matrix elements are used for purposes other than as input to TNM, a cautious user should inspect the symmetry of T_E^R to be sure that the computation is accurate enough. In particular, matrix elements with small values of k or k' may be influenced by the maximum radius at which the boundary condition is imposed, unless $kr_{\max} \gg 1$. These inaccurate values will not be a problem in most applications since they usually occur with small phase-space weighting, as is the case in TNM. As an overall indicator of the accuracy of the results of TREF, we define

$$\mathrm{SD}(l, l', J, S) \equiv \frac{\int_0^{k_{\max}} k^2 \mathrm{d}k \int_0^{k_{\max}} k'^2 \mathrm{d}k' \left[T^R_{ll'JS}(k, k') - T^R_{l'lJS}(k', k)\right]^2}{\int_0^{k_{\max}} k^2 \mathrm{d}k \int_0^{k_{\max}} k'^2 \mathrm{d}k' \left[T^R_{ll'JS}(k, k') + T^R_{l'lJS}(k', k)\right]^2}. \quad (3.27)$$

The square root of $\mathrm{SD}(l, l', J, S)$ provides a meaningful average of the likely fractional errors in the computation.

3.6.3 Input to TNM

The input to TNM is a set of files: a series (produced by TREF) containing the reference interaction and a control file whose name is given in a BLOCK DATA statement at the beginning of the main program TNM. In the version supplied, this has the default value "HEAD."

The control file ("HEAD") contains the parameters of nuclear matter, its single-particle spectrum, and the identifiers for its other input files and its output. The first record in the file is a 64-character field, which may be used as an identifier or mnemonic, and is simply passed to the output files. The second record with FORMAT[5(F8.4,1X), F10.4, 2(1X,I3),2(1X,A5)] contains, in order, the Fermi momentum p_F, the average momentum K, the maximum momentum $k_{\max}$ for use in the Legindgaard transformation (all in fm^{-1}), the starting energy in MeV, the maximum polynomial degree $n_{\max}$ (which has to be ≤ 10), the number of momenta for which the single-particle spectrum will be inputted, the identifying sequence ("TRF") for the set of files from which the reference interaction is to be obtained, and a 3-character identifying sequence for the output files. The remaining records with FORMAT[2(D16.8,1X)] each contain a value of the single-particle momentum (in fm^{-1}) and its corresponding single-particle energy (in MeV); these need to be arranged in order of increasing momentum. A sample file is included with the program for $m^*/m = 1.0$, $\gamma^2 = 1.4\ \text{fm}^{-2}$, $p_F = 1.3\ \text{fm}^{-1}$, $K = 0.7\ \text{fm}^{-1}$, $k_{\max} = 2.8\ \text{fm}^{-1}$, $E = 162.3$ MeV, $n_{\max} = 5$, and 20 points for the spectrum, which is taken to be a parabola of the form of (3.9) with $m^*/m = 1.0$ and $A = 100$ MeV; these values are chosen to give a continuous spectrum at $k_{\max}$ when used with the default values in `TREF`. The sample identifier for the output file is "SAM." A sample file is included with the program and its structure according to this description is:

Line 1				
Mnemonic				
MEMO				
'Sample for TNM control file HEAD'				
Line 2				
m^*/m	γ^2	p_F	K	$k_{\max}$
EFM	GMASQ	PFM	PAV	PKMX
1.0000	1.4000	1.3000	0.7000	2.8000
Line 2 (continued)				
E	$n_{\max}$	points	input	output
EMEV	MUP	KUP	INID	OUTID
162.3000	5	20	'TRF'	'SAM'
Line 3 etc.				
k_i	$E_{SP}(k_i)$			
PTAB(I)	ETAB(I)			
0.000	0.10000000E+03	ETAB(I)=41.47*.5*PTAB(I)**2+100.0		

The remaining input files used by `TNM` are produced by `TREF` or may be supplied by the user. There is one file in the series "TRF" for each angular-momentum channel $ll'JS$, named according to the convention $T_{ll'JS}(k,k')$ = "**TRF**JLS" for uncoupled channels, $T_{ll'JS}(k,k')$ = "**TRF**JLS" and $T_{ll'JS}(r)$ = "**TRF**JLC" for $l \neq l'$ for tensor-coupled channels. The first record in each file contains the parameters m^*/m and γ^2 (in fm^{-2}) and the number of momenta N_k at which the interaction is given, with FORMAT[2(F8.4,1X),I4]. The following N_k records with FORMAT(D16.8) each contain one value of the momentum, arranged in increasing order. The remaining $N_k(N_k+1)/2$ records each contain one pair of matrix elements,

$T^R_{ll'JS}(k,k')$ and $T^R_{ll'JS}(k',k)$ in MeV, with FORMAT[2(D16.8,1X)]; they are arranged in order of increasing k and k', with k' varying most rapidly.

3.6.4 Output from TNM

Because the output from TNM is most likely to be used as input to another computation, it consists of a series of files. For each angular-momentum channel one file is produced, named like the outputs from TRF but with the first three characters of each file name replaced by its new identifier, which was specified in the input to TNM. The first two records of each output file are copies of the first two records of the input control file ("HEAD") to TNM. The third record of each file is a copy of the first record in the corresponding angular-momentum-channel file "TRF…." The remaining $N_k(N_k+3)/2$ records are similar to those in the input, but with the nuclear-matter effective interaction instead of the reference input.

An optional series of files "INV,…," contain the results of the inverse transformations $T^R_{ll'JS}(n,n') \to T^R_{ll'JS}(k,k)$, with the Legindgaard basis truncated at $n_{\max}$, $n_{\max}-1$, and so on, along with a quantity "GSIGMA" analogous to (3.27). This check of the adequacy of the size of the polynomial basis is performed if the variable "TEST" is set ".TRUE." in the initializing DATA at the beginning of the main program.

3.7 Performance

Both codes were developed under Sun FORTRAN 1.0 on a Sun 3/60 workstation equipped with a MC68881 coprocessor. They were compatible with IBM RT/PC FORTRAN 1.0, VAX VMS FORTRAN 5.0, and HP-UX/HP-FORTRAN 77 7.0. Under VAX FORTRAN, the program TREF was compiled with the G_floating option. The source files have sizes of 53 kBytes (TREF) and 45kBytes (TNM). Typically, a calculation of a 40×40 reference reaction matrix on a position space mesh with 200 radial points took 45 s per channel. The accuracy of the matrix elements in TREF is dominated by the use of the trapezoidal rule for the radial integrals, where flexibility is accomplished at the expense of rather slow convergence. The sample mesh with 200 points yielded satisfactory accuracy and doubling the number of steps changed the matrix elements by less than 0.5%. In TNM, the best choice was a 50×50 momentum mesh; this reflects the error behavior of the two-dimensional Simpson integration in FORTRAN double precision. However, more economical sparser meshes can be employed without substantial loss in accuracy.

3.8 Things to Do

A simple application of TNM is to compute the binding energy of nuclear matter. To do this, you will first need to compute the single-particle spectrum self-consistently using the Hartree–Fock prescription for hole states with $p < p_F$, as described in [3.1,2,4,5,6].

- First, try Bethe's prescription for the single-particle energies of unoccupied states, that is, $E(p) = p^2/2m$ for $p > p_F$. How does the binding energy vary with p_F?
- Then try using a self-consistent spectrum for the unoccupied states as well. How do your results change? Compare your results with those of [3.3].
- Of course, the most interesting results will be to apply the resulting effective interaction to your favorite problem of nuclear structure. For these you will need to make a density-dependent effective interaction. For a method to construct this in coordinate representation, see [3.13].

References

[3.1] B. D. Day, Rev. Mod. Phys. **39**, 719 (1967)
[3.2] H. A. Bethe, Ann. Rev. Nucl. Sci. **21**, 93 (1971)
[3.3] J. P. Jeukenne, A. Lejeune, and C. Mahaux, Phys. Lett **C25**, 83 (1976)
[3.4] A. deShalit and H. Feshbach, *Theoretical Nuclear Physics* (Wiley, New York, 1974)
[3.5] P. J. Siemens, in *Selected Topics in Physics, Astrophysics, and Biophysics*, edited by Abecassis de Laredo and Jurisic (Reidel, Dordrecht, 1973) p. 307.
[3.6] W. Legindgaard, Nucl. Phys. **A297**, 429 (1978)
[3.7] H. A. Bethe, B. H. Brandow, and A. G. Petschek, Phys. Rev. **129**, 225 (1963)
[3.8] K. A. Brueckner and C. A. Levinson, Phys. Rev. **97**, 1344 (1955); K. A. Brueckner, Phys. Rev. **97**, 1353 (1955)
[3.9] P. J. Siemens and A. S. Jensen, *Elements of Nuclei: Many-Body Physics with the Strong Interaction* (Addison-Wesley, Reading, 1987).
[3.10] H. A. Bethe and J. Goldstone, Proc. Royal Soc. (London) **A238**, 531 (1957); J. Goldstone, Proc. Royal Soc. (London) A**293**, 267 (1957)
[3.11] B. D. Day and R. B. Wiringa, Phys. Rev. C**32**, 1057 (1985)
[3.12] R. V. Reid, Annals of Physics **50**, 411 (1968)
[3.13] P. J. Siemens, Nucl. Phys. **A141**, 225 (1970)

4. Microscopic Description of Nuclear Collisions

G. Blüge, K. Langanke, and H.-G. Reusch

4.1 Introduction

In recent years, microscopic cluster models have been applied successfully in studying light nuclear systems. These approaches are performed through many-body wave functions having well-defined quantum numbers and fulfilling the Pauli principle exactly. While the model is in principle an exact solution to the many-body problem, practical applications are restricted to the consideration of a model space spanned by a few cluster wave functions, accounting for the fact that in light nuclei, nucleons have the tendency to group in clusters. Keeping the internal degrees of freedom of these clusters fixed, the many-body problem is reduced to determining the relative motion between the various clusters. Detailed reviews on microscopic nuclear cluster models can be found in [4.1–3]. More recently, these models have also been successfully applied to light hypernuclei [4.4,5] as well as to light hadronic systems within non-relativistic quark models [4.6–8] (see also Chapt. 10).

In this chapter, we will describe a set of computer programs that allows the study of light nuclear systems in the framework of the resonating group method (RGM), which is a microscopic cluster model especially suited for few-nucleon systems at low energies where only a few channels are open. In the RGM, the clusters are described in internal coordinates, allowing the use of basis wave functions appropriate for the individual clusters and an exact treatment of the center-of-mass motion.

The main task in performing RGM calculations is the evaluation of the many-body matrix elements of the microscopic Hamiltonian and the unit operator between the various basis states, which is nowadays done on a computer with the help of algebraic-manipulation routines. However, with these methods, analytical expressions for general matrix elements can only be derived for 2-cluster basis wave functions. While for a detailed study of a nucleus, the inclusion of 3-cluster or even higher-cluster wave functions might be necessary, we will restrict ourselves in the following to a 2-cluster description of the nuclei with mass numbers $A = 5 - 8$ for which we supply routines containing analytical expressions for the various RGM integral kernels, allowing the user a rather flexible use of the programs. Furthermore, many aspects of the $A = 5 - 8$ nuclei are well-described within such an approach.

The chapter is organized as follows. In the next section, we give a brief description of the resonating group method. Some numerical aspects of the codes are discussed in Sect. 3, while Sect. 4 contains some guidelines for using the programs. Finally, in Sect. 5, we discuss some physical problems that might be studied with the help of the codes.

4.2 Theoretical Background to the Resonating Group Method

In the resonating group method, the many-body wave function is decomposed into contributions from various channels. Each channel is defined by a given number N_ν of clusters and antisymmetrized internal wave functions $\phi_1(\xi_1), \phi_2(\xi_2), \ldots, \phi_{N_\nu}(\xi_{N_\nu})$ depending on the translationally invariant internal coordinates ξ_i in the respective clusters. The internal wave functions are antisymmetrized with respect to the exchange of nucleons in the various channels.

Restricting ourselves to those cases in which all relevant channels can be described by two-cluster fragmentations, the RGM many-nucleon wave function reads

$$\psi = \sum_\nu \mathcal{A}_\nu \left[\phi_{1\nu}(\xi_{1\nu})\phi_{2\nu}(\xi_{2\nu})g_\nu(\boldsymbol{r}_\nu)\right] , \tag{4.1}$$

where $\mathcal{A}_\nu$ is the rest antisymmetrizer accounting for the exchange of nucleons between the two clusters in channel ν. The relative-motion wave function g_ν in the channel ν depends on the relative-distance coordinate

$$\boldsymbol{r}_\nu = \frac{1}{A_{1\nu}} \sum_{i\in 1} \boldsymbol{r}_i - \frac{1}{A_{2\nu}} \sum_{j\in 2} \boldsymbol{r}_j , \tag{4.2}$$

where $A_{1\nu}, A_{2\nu}$ are the numbers of nucleons in the two clusters in channel ν. Equation (4.1) can be rewritten as

$$\psi = \sum_\nu \int \mathrm{d}^3 \boldsymbol{r}_\nu \, g_\nu(\boldsymbol{r}_\nu) \, \phi_\nu(\boldsymbol{r}_\nu), \tag{4.3}$$

with

$$\phi_\nu(\boldsymbol{r}_\nu) = \mathcal{A}_\nu \left[\phi_{1\nu}(\xi_{1\nu})\phi_{2\nu}(\xi_{2\nu})\delta(\boldsymbol{r}'_\nu - \boldsymbol{r}_\nu)\right], \tag{4.4}$$

indicating that g plays the role of an expansion coefficient in the expansion of the many-body wave function ψ in terms of a basis of many-nucleon functions ϕ_ν.

The relative-motion wave functions g_ν can be determined by solving the many-body Schrödinger equation

$$(H - E)|\psi\rangle = 0 \tag{4.5}$$

in the Hilbert space spanned by the basis function (4.4).

By replacing the many-nucleon wave function ψ with the expansion (4.3) and projection onto the various channels, one obtains a set of coupled integro-differential equations, the so-called RGM equations [4.1–3]:

$$\sum_{\nu'} \int \mathrm{d}^3 \boldsymbol{r}'_{\nu'} \left[\mathcal{H}_{\nu\nu'}(\boldsymbol{r}_\nu, \boldsymbol{r}'_{\nu'}) - E\mathcal{N}_{\nu\nu'}(\boldsymbol{r}_\nu, \boldsymbol{r}'_{\nu'})\right] g_{\nu'}(\boldsymbol{r}'_{\nu'}) = 0, \tag{4.6}$$

with

$$\begin{aligned}\mathcal{N}_{\nu\nu'}(\boldsymbol{r},\boldsymbol{r}') &= \langle \phi_{1\nu}\phi_{2\nu}\delta(\boldsymbol{r}_\nu - \boldsymbol{r})|\mathcal{A}_\nu \mathcal{A}_{\nu'}|\phi_{1\nu'}\phi_{2\nu'}\delta(\boldsymbol{r}_{\nu'} - \boldsymbol{r}')\rangle \\ &= \sum_{\mathrm{P}}^{\nu}(-)^{\mathrm{P}}\langle \phi_{1\nu}\phi_{2\nu}\delta(\boldsymbol{r}_\nu - \boldsymbol{r})|P|\phi_{1\nu'}\phi_{2\nu'}\delta(\boldsymbol{r}_{\nu'} - \boldsymbol{r}')\rangle \\ \mathcal{H}_{\nu\nu'}(\boldsymbol{r},\boldsymbol{r}') &= \langle \phi_{1\nu}\phi_{2\nu}\delta(\boldsymbol{r}_\nu - \boldsymbol{r})|\mathcal{A}_\nu H \mathcal{A}_{\nu'}|\phi_{1\nu'}\phi_{2\nu'}\delta(\boldsymbol{r}_{\nu'} - \boldsymbol{r}')\rangle \\ &= \sum_{\mathrm{P}}^{\nu}(-)^{\mathrm{P}}\langle \phi_{1\nu}\phi_{2\nu}\delta(\boldsymbol{r}_\nu - \boldsymbol{r})|HP|\phi_{1\nu'}\phi_{2\nu'}\delta(\boldsymbol{r}_{\nu'} - \boldsymbol{r}')\rangle. \end{aligned} \tag{4.7}$$

Equations (4.6) have to be solved for all r and ν. In (4.7), the symbol $\sum_{\mathrm{P}}^{\nu}$ means the sum over all possible arrangements of $A = A_{1\nu} + A_{2\nu}$ nucleons into clusters of $A_{1\nu}$ and $A_{2\nu}$ nucleons, respectively. In the derivation of (4.7) we have used the facts that the rest antisymmetrizers in the various channels might be replaced by the full antisymmetrizer $\mathcal{A}$, since the internal wave functions $\phi_{i\nu}$ are antisymmetrized in themselves, and that $\mathcal{A}$ commutes with the Hamiltonian and the unit operator.

Further, H is the full microscopic Hamiltonian for which we will assume in the following that it only contains two-nucleon forces v. For $\nu = \nu'$, it is useful to decompose H into internal and relative parts

$$H = -\frac{\hbar^2}{2\mu}\Delta_{r_\nu} + V_{\mathrm{rel}}(r_\nu) + H^{\mathrm{int}}; \tag{4.8}$$

$$V_{\mathrm{rel}}(\boldsymbol{r}_\nu) = \sum_{\substack{i\in 1\\ j\in 2}} \nu_{ij}. \tag{4.9}$$

Rewriting $\sum_{\mathrm{P}}^{\nu}(-1)^{\mathrm{P}} = 1 + [\sum_{\mathrm{P}}^{\nu}(-1)^{\mathrm{P}} - 1] = 1 + \sum_{p\neq id}^{\nu}(-)^{\mathrm{P}}$, it is possible to split the integral kernels (4.7) into a local direct and a non-local exchange part:

$$\mathcal{H}_{\nu\nu}(\boldsymbol{r},\boldsymbol{r}') = \mathcal{H}_{\nu\nu}^{\mathrm{D}}(\boldsymbol{r}) + \mathcal{H}_{\nu\nu}^{\mathrm{ex}}(\boldsymbol{r},\boldsymbol{r}'), \tag{4.10}$$

$$\mathcal{N}_{\nu\nu}(\boldsymbol{r},\boldsymbol{r}') = \delta(\boldsymbol{r} - \boldsymbol{r}') + \mathcal{N}_{\nu\nu}^{\mathrm{ex}}(\boldsymbol{r},\boldsymbol{r}'), \tag{4.11}$$

$$\begin{aligned}\mathcal{H}_{\nu\nu}^{\mathrm{D}}(\boldsymbol{r},\boldsymbol{r}') &= \langle \phi_{1\nu}\phi_{2\nu}\delta(\boldsymbol{r}_\nu - \boldsymbol{r})|H|\phi_{1\nu}\phi_{2\nu}\delta(\boldsymbol{r}_\nu - \boldsymbol{r}')\rangle \\ &= -\frac{\hbar^2}{2\mu}\Delta_r + V_\nu^{\mathrm{D}}(\boldsymbol{r}) + E^{\mathrm{int}}, \end{aligned}\tag{4.12}$$

$$\mathcal{H}_{\nu\nu}^{\mathrm{ex}}(\boldsymbol{r},\boldsymbol{r}') = \sum_{\nu\neq id}(-)^{\mathrm{P}}\langle \phi_{1\nu}\phi_{2\nu}\delta(\boldsymbol{r}_\nu - \boldsymbol{r})|HP|\phi_{1\nu}\phi_{2\nu}\delta(\boldsymbol{r}_\nu - \boldsymbol{r}')\rangle, \tag{4.13}$$

$$\mathcal{N}_{\nu\nu}^{\mathrm{ex}}(\boldsymbol{r},\boldsymbol{r}') = \sum_{\nu\neq id}(-)^{\mathrm{P}}\langle \phi_{1\nu}\phi_{2\nu}\delta(\boldsymbol{r}_\nu - \boldsymbol{r})|\mathcal{N}P|\phi_{1\nu}\phi_{2\nu}\delta(\boldsymbol{r}_\nu - \boldsymbol{r}')\rangle. \tag{4.14}$$

In (4.12), E^{int} is the expectation value of the internal Hamiltonian H^{int} and V_ν^D defines the double-folding potential

$$\begin{aligned}V_\nu^{\mathrm{D}}(\boldsymbol{r}) = \sum_{i\in 1}\sum_{j\in 2}\int \mathrm{d}^3\boldsymbol{r}_1 \int \mathrm{d}^3\boldsymbol{r}_2\, &\langle \phi_{1\nu}|\delta(\boldsymbol{r}_1 - \boldsymbol{r}_{i\nu})|\phi_{1\nu}\rangle \\ &\times \langle \phi_{2\nu}|\delta(\boldsymbol{r}_2 - \boldsymbol{r}_{j\nu})|\phi_{2\nu}\rangle \nu(\boldsymbol{r} - \boldsymbol{r}_1 - \boldsymbol{r}_2), \end{aligned}\tag{4.15}$$

where we have assumed that the two-nucleon force v of (4.9) is local.

A similar decomposition of the integral kernels $\mathcal{H}$ and $\mathcal{N}$ into direct and non-local parts is possible, if for $\nu \neq \nu'$ the two channels correspond to the same arrangement of the nucleons into two clusters (for example, one of the clusters in channel ν is an excited state of a cluster in ν'). However, the product $\langle\delta(r_1 - r_{i\nu})\rangle\langle\delta(r_2 - r_{j\nu})\rangle$ has to be replaced by $\langle\phi_{1\nu}|\delta(r_1-r_{i\nu})|\phi_{1\nu'}\rangle\langle\phi_{2\nu}|\delta(r_2-r_{j\nu})|\phi_{2\nu'}\rangle$ in the definition of the double-folding potential. For rearrangement channels, $\nu \neq \nu'$, both integral kernels $\mathcal{H}_{\nu\nu'}$ and $\mathcal{N}_{\nu\nu'}$ are short-ranged and truly non-local.

The appearance of the overlap operator $\mathcal{N}_{\nu\nu'}$ on the right-hand side of (4.6) reflects the fact that in the RGM the many-body wave function ψ is expanded in terms of a non-orthogonal basis. Thus, the coupled-RGM equation does not represent a system of multichannel Schrödinger equations of relative motions. However, the non-orthogonality is short-ranged, as it originates from the exchange antisymmetrization between the nucleons of the two different clusters for a given fragmentation. Thus, asymptotically, one has

$$\mathcal{N}_{\nu\nu'}(\boldsymbol{r}, \boldsymbol{r}') \rightarrow \delta_{\nu\nu'}\delta(\boldsymbol{r} - \boldsymbol{r}'). \tag{4.16}$$

This allows us to define all physically important quantities (for example, the scattering matrix) with the asymptotic RGM wave functions $g_\nu(\boldsymbol{r})$, which obey just the same asymptotic boundary conditions as the relative wave functions in a conventional multichannel collision theory.

In practical approaches, it is useful to define the many-nucleon wave function ψ as well as the basis functions in terms of well-defined quantum numbers for the respective collision problem. In the following, we will restrict our discussion of the RGM to two-cluster wave functions with both fragments being s-shell nuclei [that is, proton, neutron, deuteron, ^{3}H (triton), ^{3}He, ^{4}He]. Furthermore, we will only consider the central, the spin-orbit, and the Coulomb part in the nucleon–nucleon interaction. Thus, the parity π, the total spin S, the total angular momentum J, its projection M on the z-axis, and the relative orbital angular momentum L are conserved quantum numbers. For a given set of quantum numbers $\{\alpha\} = \{J, M, L, S, \pi\}$, one might define the many-nucleon RGM wave function as

$$\psi_\alpha = \sum_\nu \int \mathrm{d}r_\nu r_\nu^2 \phi_{\alpha,\nu}(r_\nu) \frac{1}{r_\nu} g_\alpha(r_\nu), \tag{4.17}$$

with the basis wave functions

$$\phi_{\alpha,\nu}(r_\nu) = \mathcal{A}_\nu \left(\left\{ \left[\phi_{1\nu}^{S_1}(\xi_{1\nu})\phi_{2\nu}^{S_2}(\xi_{2\nu}) \right]^S \otimes Y_L(\hat{r}_\nu) \right\}_M^{J\pi} \frac{\delta(r_\nu - r_{\nu'})}{r_\nu'^2} \right). \tag{4.18}$$

Here, the spins of the fragments are coupled to the total spin S, which, under the present restriction, is identical to the conventional channel spin. The relative orbital angular momentum L and the channel spin S are then coupled to J, the total angular momentum. As the present paper is restricted to s-shell fragment nuclei only, the parity π is defined by the parity of the relative motion. For scattering states, the relative wave functions g_α are asymptotically linear combinations of the regular and irregular Coulomb functions $F_L(kr), G_L(kr)$:

$$g_\alpha^{\mathrm{as}} = \frac{1}{2}\,\mathrm{i}^{L+1}\,\exp\left(\mathrm{i}\sigma_L\right)\sqrt{\frac{v_{\alpha'}}{v_\alpha}} \times \left[(G_L - \mathrm{i}F_L)(k_{\alpha'}r)\delta_{\alpha'\alpha} - S_{\alpha'\alpha}(G_L + \mathrm{i}F_L)(k_\alpha r)\right], \tag{4.19}$$

where we have denoted the entrance channel by α'. The relative velocity in the channel α is v_α. The asymptotic form (4.19) ensures that the incoming wave is normalized to unit flux and that the scattering matrix, whose elements are $S_{\alpha,\alpha'}$, is unitary. Note that in (4.19) we have made use of the fact that for the problems to be studied with the present version of the code, all coupled channels have the same relative angular momentum L.

For practical purposes, it is disadvantageous to use the *complex* boundary conditions (4.19). With the help of the K-matrix transformation, it is possible to rewrite the set of RGM equations (4.6), (4.19) in a manifestly real form. The K-matrix is a real symmetric matrix which is related to the S-matrix by

$$K = \mathrm{i}\,\frac{1-S}{1+S}; \qquad S = \frac{1+\mathrm{i}K}{1-\mathrm{i}K}. \tag{4.20}$$

We then define a set of relative wave functions as

$$\tilde{g}_\alpha = \sum_{\alpha'} \mathrm{i}\,(1+S)^{-1}_{\alpha\alpha'}\,\frac{1}{c_{\alpha'}}\,g_{\alpha'}, \tag{4.21}$$

with

$$c_\alpha = \mathrm{i}^{L+1}\,\exp\left\{\mathrm{i}\sigma_L\right\}. \tag{4.22}$$

The new set of relative wave functions also has to solve the RGM equations; its asymptotic form, however, is real:

$$\tilde{g}_\alpha^{\mathrm{as}} = \frac{1}{\sqrt{v_\alpha}}\left[F_L(k_{\alpha'}r)\delta_{\alpha'\alpha} + K_{\alpha'\alpha}\,G_L(k_\alpha r)\right]. \tag{4.23}$$

For the spatial part of the various internal cluster wave functions, it has been found that these are well-approximated by Gaussian form factors. Thus, one has

$$\phi_4 = \exp\left[-\frac{1}{2}\beta_4\sum_{k=1}^{4}(\boldsymbol{r}_k - \boldsymbol{R}_4)^2\right], \boldsymbol{R}_4 = \frac{1}{4}\sum_{k=1}^{4}\boldsymbol{r}_k, \tag{4.24}$$

for the α particle,

$$\phi_3 = \exp\left[-\frac{1}{2}\beta_3\sum_{k=1}^{3}(\boldsymbol{r}_k - \boldsymbol{R}_3)^2\right], \boldsymbol{R}_3 = \frac{1}{3}\sum_{k=1}^{3}\boldsymbol{r}_k, \tag{4.25}$$

for the tritium and ^{3}He nucleus, and

$$\phi_2 = \sum_{j=1}^{3} A_j \exp\left[-\frac{1}{2}\beta_{2j}\sum_{k=1}^{2}(\boldsymbol{r}_k - \boldsymbol{R}_2)^2\right], \boldsymbol{R}_2 = \frac{1}{2}\sum_{k=1}^{2}\boldsymbol{r}_k, \tag{4.26}$$

for the deuteron, where the various parameters depend on the adopted effective interactions. In each case, they are determined to yield reasonable results for the binding energies and the rms-radius of the respective cluster nuclei. Note that a single Gaussian is not sufficient to describe the loosely bound state of the deuteron. Improved descriptions of the clusters can be obtained by adding additional Gaussian cluster functions.

Our nucleon–nucleon potential has the form

$$v_{ij} = v_{ij}^c + v_{ij}^{\text{Coul}} + v_{ij}^{ls}, \tag{4.27}$$

with

$$\begin{aligned} v_{ij}^c &= \sum_k v_k^0 \exp[-\alpha_k(\boldsymbol{r}_i - \boldsymbol{r}_j)^2](w_k + m_k P_{ij}^r + b_k P_{ij}^\sigma + h_k P_{ij}^\tau), \\ v_{ij}^{\text{Coul}} &= \frac{e^2}{|\boldsymbol{r}_i - \boldsymbol{r}_j|}\left(\frac{1}{2} + \tau_{iz}\right)\left(\frac{1}{2} + \tau_{jz}\right), \\ v_{ij}^{ls} &= -\frac{1}{2\hbar}V_\lambda \exp[-\lambda(\boldsymbol{r}_i - \boldsymbol{r}_j)^2]\left[(\boldsymbol{r}_i - \boldsymbol{r}_j)\times(\boldsymbol{p}_i - \boldsymbol{p}_j)\right]\cdot(\boldsymbol{\sigma}_i + \boldsymbol{\sigma}_j). \end{aligned} \tag{4.28}$$

Here, $P_{ij}^\sigma, P_{ij}^\tau$, and P_{ij}^r denote the usual spin, isospin, and spatial exchange operators, respectively. τ_{kz} is the z component of the isospin operator for particle k (we define $\tau_z = \frac{1}{2}$ for the proton) and σ_k is the Pauli-spin matrix for particle k. The potential parameters ($V_k^0, \alpha_k, w_k, m_k, b_k, h_k, v_\lambda$ and λ) are different for the various effective NN interactions. The parameters for three standard central interactions (Minnesota force [4.9], Hackenbroich–Heiss interaction [4,10], and Hasegawa–Nagata force [4.11]), for the spin–orbit interaction of [4.12], and the respective values for the parameters for the internal wave functions are listed at the beginning of the codes that calculate the RGM kernels.

Note that the description of s-shell clusters is not dependent on each of the parameters (w, b, h, m) individually, but rather only on a lesser number of linear combinations of these quantities. This fact is used in two of the effective NN-interactions to reduce the number of parameters and to introduce more flexibility by defining parameters of which the description of s-shell nuclei is independent, but that can be adjusted to properties of combined s-shell cluster systems. The Hasegawa–Nagata force is formulated in a way that the parameter m_2 (the Majorana exchange parameter of the Gaussian with medium range) can be interpreted as an adjustable parameter with a value phenomenologically found to be about $m_2 = 0.4$. The commonly employed version of the Minnesota potential reads

$$v_{ij}^c(r) = \frac{1}{4}\left[(1 + P_{ij}^\sigma)V_t(r) + (1 - P_{ij}^\sigma)V_s(r)\right]\left[u + (1-u)P_{ij}^r\right], \tag{4.29}$$

where V_t, V_s are the triplet and singlet potentials in even orbital angular momentum states, both given by a single Gaussian form factor. The space exchange mixture parameter u is treated as an adjustable parameter. Its value should be about $u = 1$, corresponding to a Serber mixture. A value of u noticeably larger than 1 indicates that the trial wave function does not describe the nuclear system properly [4.1]. In the Hackenbroich–Heiss force, the potential is chosen to be of the form (4.28) with the parameters satisfying the relations

$$w + m + b + h = 1, \tag{4.30}$$

$$w + m - b - h = x, \tag{4.31}$$

where x is the ratio of the s-wave singlet-to-triplet interaction. The parameter x should have the value $x = 0.63$ [4.1]. As the binding energies of s-shell nuclei (except for the deuteron) depend on x, it is dangerous to interpret x as an adjustable parameter.

4.3 A Short Description of the Numerical Methods Used and the Program Structure

The major task in performing RGM studies is the calculation of the integral kernels (4.7). Although this problem is straightforward and can be done for any reasonable basis functions, in practical applications the internal degrees of freedom are described within the harmonic-oscillator shell model for the respective nuclei. Then, the calculation can be performed analytically and the task reduces, albeit to a tedious-looking problem. In recent years, computer codes have been developed that perform this exercise on the basis of group-theoretical considerations [4.13] or by using algebraic manipulation routines like REDUCE [4.14]. Note that these rather elaborate codes can calculate the RGM kernels for two-cluster functions if both fragments are (0s)-shell nuclei described within the harmonic-oscillator shell model with different oscillator parameters for each cluster. Fujiwara et al. [4.15] have extended their code to include multicluster wave functions. However, for technical reasons, they are then restricted to the use of a common width parameter for each cluster. Assenbaum et al. [4.16,17] have developed a computer code that calculates RGM kernels for 3-cluster wave functions numerically, allowing for different width parameters of each fragment.

We have determined analytical expressions for the integral kernels (4.7) for various two-cluster configurations of s-shell nuclei adopting the Hamiltonian (4.8,9). These kernels can be found in a separate file for each combination of fragmentation presently considered; a listing of these codes is given in Table 4.1. Running these codes will create data files containing the respective integral kernels on a mesh with equal spacing. These data files then have to be used as input to the program KMAT, which solves the coupled RGM equations using the Robertson method [4.18], which is basically a generalization of the Numerov algorithm to integro-differential equations. The output of the program is the scattering matrix and, optionally, the wave function of relative motion. To fit on a PC, we have restricted the coupled-channel to two-channel problems only. However, it is straightforward to extend it to more channels by changing some of the array dimensions. As the code involves a matrix inversion, it is not suitable if the number of channels becomes too large. We have found that it is capable of solving a system of about 10 coupled equations. For a significantly larger number of channels, it might be better to use an algorithm based on a variational procedure (see Chapt. 10).

Table 4.1. Listing of the codes that calculate the RGM kernels for the various 2-cluster fragmentations

Code name	Fragmentation
`NALPHA`	n+α, p+α
`DHEDT`	d+^{3}H, d+^{3}He
`COUPLE`	coupling between (d+^{3}He,p+α), (d+^{3}H,n+α)
`ALPHAD`	d+α, (np)+α
`ALPHATRI`	^{3}H+α, ^{3}He+α
`ALPHALPH`	α+α

The resonating group method provides a consistent description of scattering and bound states. Owing to array restrictions, the present PC-code allows only a direct solution of the collision problem, while for the determination of bound states, however, some rather simple modifications are required, which we will now briefly discuss.

The easiest way to calculate bound states of the two-cluster systems studied in this chapter is by determining the energy eigenvalues of the matrix relation $\mathcal{H} - E\mathcal{N} = 0$, which, for example, can be performed by using the routine `F02AEF` of the NAG library. The required RGM Hamiltonian and norm kernels can be readily obtained from the respective files, listed in Table 4.1. Note that in the procedure suggested above, we simply use the boundary condition that the relative wave function is zero outside a certain matching radius rather than matching the relative wave function to the appropriate Whittaker function. This simplification is not valid for weakly bound states, while it is justified for the bound states of the two-cluster-systems studied here.

4.4 Installation and Use of the Codes

Before starting the actual RGM calculations, one should compile the codes contained in the file `HILF.FOR`. As this file is a library of programs required by the codes listed in Table 4.1 and by the RGM equation solver `KMAT`, the latter have to be linked to `HILF` before execution. Additionally, some programs from the NAG library are required (see below).

As a first step for performing an RGM calculation, one has to run the codes that generate the discretized RGM kernels needed for the study (see Table 4.1). This is a single code for nuclear systems with only one 2-cluster fragmentation, while for an investigation of the d(^{3}He,p)α or d(^{3}H,n)α reactions, it is necessary to run the three codes `NALPHA.FOR`, `DHEDT.FOR`, and `COUPLE.FOR`. For the latter use in the RGM equation solver `KMAT`, we have adopted the following conventions: The name of the file generated by `COUPLE` should start with a '+' sign (e.g. '+coupl.out') if one wants to study the d(^{3}He,p)α or the d(^{3}H,n)α reaction. In this case, the file generated by `DHEDT` is in in the upper left-hand corner of the RGM matrices H and N. If one is interested in the inverse reactions, the file generated by `COUPLE` should begin with a '−' sign. In this case, the coupling matrix will be automatically transposed in `KMAT`. If the name of the file containing the coupling matrices will start with '0', coupling of the channels will be ignored in `KMAT`.

The general input to the codes listed in Table 4.1 runs as follows:

1. FRAG – charge number (integer) of the lighter fragment
   This quantity allows one to distinguish between the two different fragmentations for $T = 1/2$ systems. It is not needed in `ALPHAD` and `ALPHALPH`.

2. PO – choice of central NN potential (integer)
   - PO=1: Minnesota potential
   - PO=2: Hackenbroich–Heiss potential
   - PO=3: Hasegawa–Nagata potential

3. DIS – number of distortion channels
   This quantity allows one to consider specific distortion effects of the deuteron following the pseudo-state method of Tang et al. [4.19]. For the present deuteron wave function (4.26), the maximum number of distortion channels is restricted to DIS=3. For DIS=1, only the (physical) bound state of the deuteron is considered. For DIS=2 (3), the program includes the (unphysical) first excited (both excited) level. This input is not required in `NALPHA.FOR`, `ALPHTRI.FOR`, and `ALPHALPH.FOR`.

4. JL,L,SBRA,SKET – integers to define the angular-momentum quantum numbers
   JL,L,SBRA,SKET determine the total angular momentum of the nuclear system, the orbital angular momentum of the relative wave function, and the spins of the configurations in the bra and in the ket of the many-body matrix elements. If the spin of the system is unique (this is the case for all fragmentations except for d+^{3}H and d+^{3}He), no input

for SKET is required. Note that the input for JL,SBRA,SKET is twice the physical quantum numbers.

5. MM – adjustable parameter in the NN potential
These parameters are discussed in Sect. 4.3. Reasonable values are given in parentheses.
   - u [1.0]: Minnesota potential
   - x [0.63]: Hackenbroich–Heiss potential
   - m_2 [0.40]: Hasegawa–Nagata potential

6. VO,VT,LAM – parameters in the spin–orbit potential
Note that the spin-orbit potential often used to study the seven-nucleon system has a slightly different form than defined in (4.28). In this case, the form factor V_λ has to be replaced by $V_0 + P_{ij}^{\tau} V_\tau$, where the respective parameter values can be found in the file `ALPHATRI.INP`. For the spin-orbit potential (4.28), the input line simply reads VL,LAM (see `NALPHA.INP`).

7. HSTEP,STSANZ – numerical parameters
   - HSTEP: step size
   - STSANZ: number of mesh points

For each of the codes listed in Table 4.1, we provide a sample input file. For convenience of the user, these files are commented.

After all required discretized kernels have been calculated and are written on files, one can solve the RGM equations (4.6). This task is done in the program `KMAT.FOR`. To run the code one needs the input file `KMAT.INP`; an example file is included on the disk. Although this example file is well-commented and self-explanatory, we will in the following give a listing of the required input. Note that for the sake of clarity, the input is formatted and contains some empty lines, which we will skip in our listing.

1. ZTOTJ,NTOTJ – nominator and denominator defining the total angular momentum (for example, 3/2, 2/1).

2. K – number of channels (integer)

3. H0,NMAX – minimal step size, maximum number of mesh points

4. HSTEP,NS – actual step size and number of mesh points (integer)
Note that the inequality (NS+1)*HSTEP $\leq$ (NMAX+1)*H0 has to be fulfilled.

   Following a line of comment, which for the convenience of the user orders the numbers of channels, one has to define

5. AMUE(i) – the reduced mass in each channel i

6. NZ(i) – the product of charge numbers $Z_{i1} * Z_{i2}$ in each channel i (integer)

   Again, following a line of comment, the program expects input for

7. CHSPIN(i) – spin-statistical factor $(2I_{i1}+1)*(2I_{i2}+1)$ in each channel i where I_1, I_2 are the spins of the two respective fragments

8. LL(i) – relative orbital angular momentum in each channel i (integer)

9. `INPNAME(i,j)` – name of files from which the discretized matrix elements are to be read

   The indices i,j run over all channels 1, ..., k.

10. MAU – output of wave function?
    - if MAU=1, wave functions are written on file
    - if MAU≠1, no output of wave functions

11. NUMBER – number of files necessary for output of wave functions

12. `NBUNAME` – name of files on which wave functions should be written

13. EANF – starting energy

14. ESTEP,EANZ – energy step size and number of energy points

The output of the program `KMAT` is written on the file `KMAT.OUT`; a representative example file is included on the disk. Besides some statistics defining the studied problem, the code gives out the elements of the complex scattering matrix in absolute value and phase at each of the required energy values and tests the unitarity of the S-matrix. Additionally, the partial cross section

$$\sigma^J(E) = \frac{\pi}{k^2} \sum \frac{2J_\alpha + 1}{(2I_1+1)(2I_2+1)} |\delta_{1\alpha} - S_{1\alpha}|^2 \tag{4.32}$$

is calculated, where I_1, I_2 are the spins of the respective fragments in the entrance channel. At energies far below the Coulomb barrier, it is more convenient to define the cross section in terms of the astrophysical S-factor

$$S(E) = E\,\sigma(E)\,\exp(2\pi\eta)\,, \tag{4.33}$$

$$2\pi\eta = 0.989\,Z_1\,Z_2\,\sqrt{\mu/E}, \tag{4.34}$$

where Z_1, Z_2 are the charge numbers of the fragments in the entrance channel and μ is its reduced mass. In (4.33,34), the energy is defined in MeV.

The (binary) output of the relative wave functions is optional, see above.

4.5 Physical Applications

To become familiar with the code and the physics behind the cluster models, it is reasonable to start with an investigation of the low-energy $p+\alpha$ or $n+\alpha$ system. Note that there are no bound states in the $A = 5$ nuclei (correspondingly in the RGM description of ^{5}Li and ^{5}He). Furthermore, the first particle channel opens at $Q = 18.35$ MeV in ^{5}Li (this is the d+^{3}He channel) and at $Q = 17.59$ MeV in ^{5}He (the $d+t$ channel). Thus, it turns out that at energies $E < Q$, the 5-nucleon-systems are well-described within a one-channel RGM calculation in the framework of the present code, considering only a $p+\alpha$ and $n+\alpha$ fragmentation, respectively. The reader will find detailed descriptions of such studies in [4.1,12].

In a first step, one might perform a calculation using only the central part in the effective interaction (that is, adopting the Minnesota potential [4.9] with $u = 0.95$) and the Coulomb potential, thus neglecting the spin–orbit force. As for $E < Q$, the phase shifts are rather small for larger l-values, and the calculation can be restricted to $l \leq 2$. Compared with the experimental phase shifts [4.20,21], one finds a good agreement in the s-wave, while the p-wave shifts lie between the experimental values for the $J^\pi = 3/2^-$ and $1/2^-$ partial waves, owing to the neglect of the spin–orbit interaction. One observes that the p-wave phase shifts are resonant while the s-wave phase shifts decrease with energy, corresponding well to hard-sphere scattering.

It is illuminating to understand why the lowest state in the 5-nucleon system occurs in the p-wave rather than in the s-wave, although the latter is favored by the centrifugal potential. This is due to the Pauli principle. The s-wave function of relative motion $g_{l=0}(r)$ is not allowed to have a contribution from the energetically most favored spherical-harmonic-oscillator state $u_{N=0,l=0}(\beta, r)$ (where N is the main quantum number and the width β is related to the width b of the internal α-particle wave function by $\beta = b\sqrt{\mu}$, μ being the relative mass number) as this would violate the Pauli principle, that is, the Pauli principle forbids bringing a fifth nucleon into the s-shell of a nuclear-shell model. A wave function that is excluded from the many-body relative-motion function by the Pauli principle is called a Pauli forbidden state [4.22]. For the 5-nucleon system, $u_{N=0,l=0}$ is the only Pauli forbidden state, as no restrictions due to the Pauli principle apply to the other partial waves $l > 0$. Note that the requirement of orthogonality to Pauli forbidden states (similar to bound states) introduces a nodal structure into the relative motion function at separations that are small compared with the nodes in the Coulomb waves (the repulsive effect of the Pauli principle is relatively short-ranged [4.23]). For a quantitative insight, the reader might compare the calculated relative-motion functions in the s- and p-waves (the latter will also have a node at energies above the resonance).

In a second step, one might improve the calculation by also considering the spin–orbit interaction [4.1,12], which results in a noticeable splitting of the p-wave phase shifts. The resonances found in the p-waves with $J^\pi = 3/2^-$ correspond to the ground states of ^{5}Li and ^{5}He, respectively, while those in the $J^\pi = 1/2^-$ partial waves describe the first excited states of these nuclei. The widths of the resonances might be estimated from the phase shifts by

$$\Gamma = 2\left(\frac{\mathrm{d}\delta}{\mathrm{d}E}\right)^{-1}_{E=E_\mathrm{R}},$$

where the resonance energy E_R can be identified with the energy where the derivative is largest [4.24]. Thus, for broad resonances, E_R will in general not coincide with the energy where the phase shifts pass through $\pi/2$.

For studies of the 5-nucleon system at energies $E > Q$, it is necessary to consider d+^{3}He and $d+t$ fragmentations, respectively. The spins of these clusters can be coupled to $S = 1/2$ and $3/2$, while in our studies the $p+\alpha$ and $n+\alpha$ fragmentations have $S = 1/2$. Thus, a coupling of these fragmentations to the d+^{3}He ($d+t$) channel is not possible via the central component in the NN interaction. Thus, one might study the $S = 3/2$ channels in a one-channel approach considering only the d+^{3}He ($d+t$) fragmentation. Such a calculation reveals a well-pronounced resonance in the s-wave phase shifts corresponding to the second excited states in ^{5}Li and ^{5}He, respectively [4.9]. For a quantitative reproduction of this state, it might be necessary to use a slightly different value for the parameter u than in the studies of the $p+\alpha$ and $n+\alpha$ systems. In the $l = 2$ partial wave, the phase shifts exhibit a broad resonance that is split into states with $J^\pi = 7/2^+, 5/2^+, 3/2^+$, and $1/2^+$ if the spin–orbit force is considered. The p-wave scattering is nonresonant. It is worthwhile reconsidering the role of the Pauli principle for this channel considering that the many-body wave function must have $S = 3/2$.

For $S = 1/2$, one might perform two-channel studies of the d+^{3}He ($d+t$) and $p+\alpha$ ($n+\alpha$) fragmentations that can couple via the central and spin–orbit components in the NN interaction. Reasonable things to investigate are: (i) How do the calculated d+^{3}He phase shifts compare with the experimental ones? (ii) Are there additional states due to the d+^{3}He (d+t) fragmentations? (iii) How strongly are the p-wave resonances affected by coupling to the closed d+^{3}He (d+t) channels? Results of a two-channel RGM calculation considering d+^{3}He and $p+\alpha$ fragmentations are reported in [4.25].

Note that the low-energy transfer reactions $d(^3\mathrm{He},p)\alpha$ and $d(t,n)\alpha$ are strongly dominated by the $J^\pi = 3/2^+$ resonance, which owing to its spin $S = 3/2$ can only couple to the α-channel via the tensor component in the NN interaction, which is missing in the present version of the code (see, however, [4.26]). The reader might find further motivation in [4.27,28]. It is easy to modify the present code to allow for the considerations of deuteron pseudo-state (see [4.27]) which might simulate the breakup of the deuteron. However, the code requires then the consideration of more than two channels and does not fit on a PC.

At low energies, the nuclei ^{7}Li, ^{7}Be, and ^{8}Be can be well-studied in one-channel approaches on the basis of $t+\alpha$, ^{3}He + α, and $\alpha+\alpha$ fragmentations, respectively, which describe the bound-state spectrum (there are no bound states in the ^{8}Be) as well as the scattering phase shifts. Reports on RGM studies of these nuclei can be found in [4.1,29,30]. More detailed investigations require the consideration of 3-cluster and pseudo-state configurations [4.13,14,16,31].

The low-energy $T = 0$ spectrum of ^{6}Li as well as low-energy $d + \alpha$ scattering are well-described within a one-channel calculation considering only a $d + \alpha$ fragmentation [4.32]. In this model space, the ground state of ^{6}Li with $J^\pi = 1^+$ corresponds to a bound d + α state in a relative s-wave. The first three excited states are reasonably well-described as resonances in the partial wave $l = 2$, split by the spin–orbit interaction. The description is improved if pseudo-deuteron states are considered [4.19]. A measure of the importance of additional configurations can be had by calculating the gain in binding energy of the ground state or in the resonance energy of narrow resonances. What is the result for the ^{6}Li ground state or for the narrow 3^+ resonance if deuteron pseudo-states are added to the model space? However, for an even more realistic description of the spectrum of ^{6}Li, including $T = 0$ and $T = 1$ states, it is necessary to include additional (^{5}Li+n, ^{5}He+p) and ^{3}He+t configurations [4.33].

4.6 Technical Note

The code mainly follows the FORTRAN 77 standard, with the exception that it uses complex*16 arithmetic. A detailed description of the code, of its implementation and use, as well as of its input and output is given in Sect. 4.4. As the codes, which create the files containing the discretized RGM kernels, are rather time-consuming, it is recommended to run them on a workstation. The code calls some NAG subroutines. These are the routines `F03AFF`, `F03AHF`, `F04AJF`, and `F04AKF` in the program `KMAT`, while the subroutines `D01BAF` and `S15AEF` are called in the library file `HILF`. In the following, we give a brief description of these NAG subroutines.

S15AEF

Purpose:		To calculate the value of the error function erf(x) via the routine name
Usage:		`REAL FUNCTION S15AEF(X,IFAIL)`
Arguments:		
X	–	Specifies the argument x (input)
IFAIL	–	integer for error detection

D01BAF

Purpose:		To compute the definite integral of a function of known analytical form, using a Gaussian quadrature formula with a specified number of abscissae
Usage:		`REAL FUNCTION D01BAF(D01BAZ,A,B,N,FUN,IFAIL)`
Arguments:		
D01BAZ	–	NAG routine for Gauss–Legendre quadrature on a finite interval
A	–	the lower limit of the integral (input)
B	–	the upper limit of the integral (input)
N	–	number of abscissae (input)
FUN	–	integrand in analytical form (input) must be declared as EXTERNAL in the calling program
IFAIL	–	integer for error detection

F03AFF

Purpose:		To decompose a real matrix A into triangular matrices using Crout's method, that is A=LU, and to determine the determinant of A
Usage:		`CALL F03AFF (N,EPS,A,IA,D1,ID,P,IFAIL)`
Arguments:		
N	–	Order of matrix A (input)
EPS	–	Specifies the smallest positive number on the computer, such as 1.0 + EPS > 1.0 (input)
A	–	Real array of dimension (IA,R), where R≥N (input) Before entry, A contains the elements of the matrix. On exit, the lower triangle contains L and the strict upper triangle contains the elements of U
IA	–	First dimension of array A (input)
D1	–	Contains part of the determinant of A such that DET = D1*2.0*ID
ID	–	Contains part of the determinant of A such that DET = D1*2.0*ID
P	–	Contains N integers such that the Ith element gives the number of the Ith pivotal row
IFAIL	–	integer for error detection

F03AHF

Purpose:		To decompose a complex matrix A into triangular matrices using Crout's method, that is A=LU, and to determine the determinant of A
Usage:		`CALL F03AHF (N,A,IA,DETR,DETI,ID,RINT,IFAIL)`
Arguments:		
N	–	Order of matrix A (input)
A	–	Complex array of dimension (IA,R), R≥N (input) Before entry, A contains the elements of the matrix. On exit, the lower triangle contains L and the strict upper triangle contains the elements of U.
IA	–	First dimension of array A (input)
DETR	–	Contains part of the determinant of A such that DET = (DETR + i*DETI)*2.0*ID
DETI	–	Contains part of the determinant of A such that DET = (DETR + i*DETI)*2.0*ID
ID	–	Contains part of the determinant of A such that DET = (DETR + i*DETI)*2.0*ID
RINT	–	Contains N integers such that the Ith element gives the number of the Ith pivotal row
IFAIL	–	integer for error detection

F04AJF

Purpose:		To calculate the approximate solution of a set of real linear equations with multiple right-hand sides, AX=B, where A has been decomposed into triangular matrices using the NAG routine `F03AFF`
Usage:		`CALL F04AJF (N,IR,A,IA,P,B,IB)`
Arguments:		
N	–	Order of matrix A (input)
IR	–	Specifies the number of right-hand sides (input)
A	–	Real array of dimension (IA,R), where R≥N (input) Before entry, A must contain the Crout factorization, A=LU, as given by `F03AFF`
IA	–	First dimension of array A, IA≥N (input)
P	–	Real array of dimension at least N (input) containing the details of the row-interchanges in Crout factorization as given by `F03AFF`
B	–	Real array of dimension (IB,Q), where Q≥IR (input) Before entry, B contains the IR right-hand sides. On exit, it contains the IR solution vectors.
IB	–	First dimension of array B, IB≥N (input)

F04AKF

Purpose:		To calculate the approximate solution of a set of complex linear equations with multiple right-hand sides, AX=B, where A has been decomposed into triangular matrices using the NAG routine `F03AHF`
Usage:		`CALL F04AKF (N,IR,A,IA,P,B,IB)`
Arguments:		
N	–	Order of matrix A (input)
IR	–	Specifies the number of right-hand sides (input)
A	–	Complex array of dimension (IA,R), R≥N (input) Before entry, A must contain the Crout factorization, A=LU, as given by `F03AHF`
IA	–	First dimension of array A, IA≥N (input)
P	–	Real array of dimension at least N (input) containing the details of the row-interchanges in Crout factorization as given by `F03AHF`
B	–	Complex array of dimension (IB,Q), Q≥IR (input) Before entry, B contains the IR right hand sides. On exit, it contains the IR solution vectors.
IB	–	First dimension of array B, IB≥N (input)

References

[4.1] K. Wildermuth and Y.C. Tang, *A Unified Theory of the Nucleus* (Vieweg, Braunschweig, 1977)
[4.2] Y.C. Tang, M. LeMere, and D.R. Thompson, Phys. Rep. **47**, 167 (1978)
[4.3] K. Langanke and H. Friedrich, *Advances in Nuclear Physics*, eds. J.W. Negele and Erich Vogt (Plenum, New York, 1987)
[4.4] Progr. Theor. Phys. Suppl. **81** (1985)
[4.5] T. Warmann and K. Langanke, Z. Phys. **A334**, 221 (1989)
[4.6] N. Isgur and G. Karl, Phys. Rev. **D19**, 2653 (1979)
[4.7] A. Faessler, F. Fernandez, G. Lübeck, and K. Shimizu, Phys. Lett. **112B**, 201 (1982)
[4.8] M. Oka and K. Yazaki, Progr. Theor. Phys. **66**, 551 and 572 (1981)
[4.9] F.S. Chwieroth, R.E. Brown, Y.C. Tang, and D.R. Thompson, Phys. Rev. C**8**, 938 (1973)
[4.10] P. Heiss and H.H. Hackenbroich, Phys. Lett. **30B**, 373 (1969)
[4.11] F. Tanabe, A. Tohsaki, and R. Tamagaki, Progr. Theor. Phys. **53**, 677 (1975)
[4.12] I. Reichstein and Y.C. Tang, Nucl. Phys. **A158**, 529 (1970)
[4.13] Y. Fujiwara and Y.C. Tang, Phys. Rev. **C31**, 342 (1985)
[4.14] G. Blüge, doctoral thesis (Münster, 1990, unpublished)
[4.15] Y. Fujiwara, Q.K.K. Liu, and Y.C. Tang, Phys. Rev. C**38**, 1531 (1988)
[4.16] H.J. Assenbaum, doctoral thesis (Münster, 1987, unpublished)
[4.17] E. Kolbe, K. Langanke, and H.J. Assenbaum, Phys. Lett. B**214**, 169 (1988)
[4.18] H.H. Robertson, Cambridge Phil. Soc. **52**, 538 (1956)
[4.19] H. Kanada, T. Kaneko, S. Saito, and Y.C. Tang, Nucl. Phys. A**444**, 209 (1985)

[4.20] P. Schwandt, T.B. Clegg, and W. Haeberli, Nucl. Phys. **A163**, 432 (1971)
[4.21] A. Houdayer, N.E. Davison, S.A. Elbakr, A.M. Sourkes, W.T.H. van Oers, and A.D. Bacher, Phys. Rev. C**18**, 1985 (1978)
[4.22] S. Saito, Progr. Theor. Phys. **41**, 705 (1969); S. Saito, S. Okai, R. Tamagaki and M. Yasuno, Progr. Theor. Phys. **50**, 1561 (1973)
[4.23] H. Friedrich and K. Langanke, Phys. Rev. C**28**, 1385 (1983)
[4.24] K.Langanke and D. Frekers, Nucl. Phys. A**302**, 134 (1978)
[4.25] F.S. Chwieroth, Y.C. Tang, and D.R. Thompson, Phys. Rev. C**9**, 56 (1974)
[4.26] G. Blüge and K. Langanke, Phys. Rev. C**41**, 1191 (1990)
[4.27] P.N. Shen, Y.C. Tang, Y. Fujiwara, and H. Kanada, Phys. Rev. C**31**, 2001 (1985)
[4.28] G. Blüge and K. Langanke, Few-Body Systems **11**, 137 (1991)
[4.29] R.D. Furber, doctoral thesis, (Minneapolis, 1975, unpublished)
[4.30] H. Federsel, E.J. Kanellopoulos, W. Sünkel, and K. Wildermuth, Phys. Lett. B**33**, 140 (1970)
[4.31] H.M. Hoffmann, T. Mertelmeier, and W. Zahn, Nucl. Phys. A**410**, 208 (1983)
[4.32] D.R. Thompson and Y.C. Tang, Phys. Rev. C**8**, 1649 (1973)
[4.33] H. Kanada, T. Kaneko, and Y.C. Tang, Phys. Rev. C**38**, 2013 (1988)

5. The Distorted-Wave Born Approximation

P.D. Kunz and E. Rost

5.1 General Description of DWUCK4

The computer code DWUCK4 calculates the scattering and reaction observables for binary nuclear reactions using the distorted-wave Born approximation (DWBA). The calculations are performed using a zero-range interaction. This restriction is, in general, not important for inelastic scattering; however, it is an approximation for the calculation of particle transfer reactions. The spin of the projectiles may be any combination of spin 0, spin 1/2, or spin 1. The following physical description uses the notation and formulation of [5.1], which describes nuclear reaction theory in detail.

The computer code calculates a transition amplitude for the reaction $A(a,b)B$ of the form

$$T = J \int \mathrm{d}^3 r_b \int \mathrm{d}^3 r_a \chi^{(-)}(\boldsymbol{k}_\mathbf{f}, \boldsymbol{r}_\mathbf{b})^* \, \langle bB|V|aA\rangle \, \chi^{(+)}(\boldsymbol{k}_\mathbf{i}, \boldsymbol{r}_\mathbf{a}), \tag{5.1}$$

where $\chi^{(-)}$ and $\chi^{(+)}$ are the distorted waves, $\boldsymbol{r}_\mathbf{a}$ and $\boldsymbol{r}_\mathbf{b}$ are the relative coordinates for the systems (a, A) and (b, B), respectively, and J is the Jacobian for the transformation to these coordinates. The quantity $\langle bB|V|aA\rangle$ is the form factor for the reaction and must contain a delta function between the coordinates $\boldsymbol{r}_\mathbf{a}$ and $\boldsymbol{r}_\mathbf{b}$. The distorted waves $\chi^{(\pm)}(\boldsymbol{k}, \boldsymbol{r})$ asymptotically describe a plane wave of momentum $\boldsymbol{k}$ plus an outgoing (or incoming) spherical scattered wave, which in the case of no Coulomb potential has the form

$$\chi^{(\pm)}(\boldsymbol{k}, \boldsymbol{r}) \longrightarrow \mathrm{e}^{\mathrm{i}\boldsymbol{k}\cdot\boldsymbol{r}} + f(\theta)\frac{\mathrm{e}^{\pm \mathrm{i}kr}}{r}. \tag{5.2}$$

The final distorted wave, which has an incoming scattered wave condition, is related to the solution with the outgoing waves by

$$\chi^{(-)*}(\boldsymbol{k}, \boldsymbol{r}) = \chi^{(+)}(-\boldsymbol{k}, \boldsymbol{r}). \tag{5.3}$$

When the initial and final projectiles have spin, the distorted waves are matrices in spin space,

$$\chi^{(\pm)}(\boldsymbol{k}, \boldsymbol{r})\eta_{s,m} = \sum_{m'} \chi^{(\pm)}_{m,m'}(\boldsymbol{k}, \boldsymbol{r})\eta_{s,m'}, \tag{5.4}$$

where the $\eta_{s,m}$ are spin functions,

$$\begin{aligned} \chi^{(\pm)}_{m,m'}(\boldsymbol{k}, \boldsymbol{r}) &= \frac{\sqrt{4\pi}}{kr} \sum_{J,L} \mathrm{i}^L \sqrt{2L+1}\, \chi_{JLs}(k, r)\, (L\, s\, M\, m | J\, M') \\ &\quad \times (L\, s\, M' - m'\, m' | J\, M')\, Y_L^{M'-m'}(\hat{\mathbf{r}}) d^L_{0,M'-m'}(\hat{\mathbf{k}}), \end{aligned} \tag{5.5}$$

and the $d^L_{0,m}$ are the rotation functions for integer spin [5.2]. The radial part of the distorted waves satisfies the equation

$$\left\{\frac{\mathrm{d}^2}{\mathrm{d}r^2}+k^2-\frac{L(L+1)}{r^2}-\frac{2\mu}{\hbar^2}[U(r)+U_{\mathrm{c}}(r)+U_{Ls}(r)\mathbf{L}.\mathbf{s}]\right\}\chi_{JLs}(k,r)=0. \tag{5.6}$$

Here, U is a central potential with real and imaginary parts, U_{c} is a Coulomb potential for a uniform charge distribution of radius R_{c}, and U_{Ls} is a spin–orbit potential. The radial functions $\chi_{JLs}(k,r)$ satisfy the boundary conditions $\chi_{JLs}(k,0)=0$ at the origin and

$$\chi_{JLs}(k,r)\longrightarrow\frac{\mathrm{i}}{2}[H_L^-(kr)-\eta_L^J H_L^+(kr)]\,\mathrm{e}^{\mathrm{i}\sigma_L} \tag{5.7}$$

for large r (where U and U_{Ls} can be neglected). Here, $H_L^{\pm}(kr)=G_L(kr)\pm iF_L(kr)$ are the outgoing (+) and the incoming (−) Coulomb waves, η_L^J is the elastic scattering S-matrix, and σ_L is the Coulomb phase shift.

The form factor $\langle bB|V|aA\rangle$ contains the nuclear-structure information and is expressed by

$$\begin{aligned}&\langle J_BM_Bs_bm_b|V|J_AM_As_am_a\rangle=\textstyle\sum_{lsj}(J_A\,j\,M_A\,M_B-M_A|J_B\,M_B)\\&\times(s_a\,s\,m_b\,m_a-m_b|s_a\,m_a)(l\,s\,m\,m_a-m_b|j\,M_A-M_B)\\&\times B_{lsj}\;f_{lsj}(r_c)\delta(r_b-\tfrac{A}{B}r_a)\mathrm{i}^{-l}Y_l^m(\hat{r}_a)^*.\end{aligned} \tag{5.8}$$

The quantity B_{lsj} is a measure of the strength of the interaction and is equal to $\sqrt{(2s+1)/(2s_a+1)}A_{lsj}$ of [5.1]. The function f_{lsj} is the radial form factor for the reaction. The details of the reaction model are contained in f_{lsj} and are discussed in Sect. 5.2 for typical cases. In the above equation, the angular momenta must satisfy the triangular relations

$$\boldsymbol{j}=\boldsymbol{J}_{\mathrm{B}}-\boldsymbol{J}_{\mathrm{A}},\qquad \boldsymbol{s}=\boldsymbol{s}_{\mathrm{a}}-\boldsymbol{s}_{\mathrm{b}},\qquad \boldsymbol{l}=\boldsymbol{j}-\boldsymbol{s}. \tag{5.9}$$

With the definitions of $\langle bB|V|aA\rangle$, and the distorted wave functions $\chi_{m,m'}^{(\pm)}$, we can write down the transition amplitude,

$$\begin{aligned}T^{M_AM_B;m_am_b}&=\frac{\sqrt{4\pi}}{k_ak_b}\sum_{l,s,j}\sqrt{2l+1}\,B_{lsj}\\&\quad\times(J_A\,j\,M_A\,M_B-M_A|J_B\,M_B)S_{lsj}^{mm_am_b},\end{aligned} \tag{5.10}$$

where the angle-dependent amplitude S is written as

$$S_{lsj}^{mm_am_b}=\sum_{L_b}\beta_{lsj;L_b}^{mm_am_b}\,P_{L_b}^{m_a-m-m_b}. \tag{5.11}$$

The inelastic scattering amplitudes β are given by

$$\begin{aligned}\beta_{lsj;L_b}^{mm_am_b}&=\sum_{J_aL_aJ_bm}(L_a\,s_a\,0\,m_a|J_a\,m_a)(L_b\,s_b\,m_a-m-m_b\,m_b|J_b\,m_a-m)\\&\quad\times(J_b\,j\,m_a-m\,m|J_a\,m_a)(2L_b+1)(L_b\,l\,0\,0|L_a\,0)\\&\quad\times\sqrt{(2s_a+1)(2j+1)(2J_b+1)(2L_a+1)}\begin{Bmatrix}L_b&s_b&J_b\\l&s&j\\L_a&s_a&J_a\end{Bmatrix}\\&\quad\times I_{J_aL_aJ_bL_b}^{lsj}\,\mathrm{i}^{L_a-L_b+l}.\end{aligned} \tag{5.12}$$

In the above expression the usual nine-j symbol, {} [5.2], appears. The radial integrals I are

$$I^{lsj}_{J_a l_a J_b l_b} = \frac{CB}{A^2} \int_0^\infty \mathrm{d}r_c \, \chi^{(-)}_{J_b L_b}\left(k_b, \frac{A}{B} r_a\right) f_{lsj}(r_c) \, \chi^{(+)}_{J_a L_a}(k_a, r_a), \tag{5.13}$$

where C is the mass of the form factor core. The radial integrals I and the scattering amplitudes β may be printed in the output (see Sect. 5.3 below).

The differential cross section for the reaction $A(a,b)B$ may now be expressed in terms of the transition amplitude T:

$$\begin{aligned} \frac{\mathrm{d}\sigma(\theta)}{\mathrm{d}\Omega} &= \left(\frac{\mu_b}{2\pi\hbar^2}\right)^2 \frac{v_b}{v_a} \frac{1}{(2J_A+1)(2s_a-1)} \sum_{M_A M_B m_a m_b} |T^{M_A M_B; m_a m_b}|^2 \\ &= \frac{1}{4\pi} \frac{2J_B+1}{2J_A+1} \frac{1}{E_a E_b} \frac{k_b}{k_a} \frac{1}{2s_a+1} \\ &\times \sum_{m_a m_b m} |\sum_{lsj} \sqrt{2l+1} B_{lsj} S^{m m_a m_b}_{lsj}|^2, \end{aligned} \tag{5.14}$$

where E_a and E_b are the center-of-mass energies for the entrance and exit channels, respectively.

The computer program DWUCK4 computes the cross section with different normalizations depending upon whether the case being run is an inelastic scattering or a particle transfer reaction. In the case of inelastic excitations (where the mass of a and b are equal), we define

$$\sigma^{lsj}_{DW}(\theta) = \frac{1}{4\pi} \frac{1}{E_a E_b} \frac{k_b}{k_a} \frac{1}{2s_a+1} \sum_{m_a m_b m} |\sum_{lsj} S^{m m_a m_b}_{lsj}|^2. \tag{5.15}$$

For the case of particle transfer (where the masses a and b are unequal), we define

$$\sigma^{lsj}_{DW}(\theta) = \frac{1}{4\pi} \frac{1}{E_a E_b} \frac{k_b}{k_a} \frac{10^4}{2s_a+1} \sum_{m_a m_b m} |\sum_{lsj} \sqrt{2l+1} S^{m m_a m_b}_{lsj}|^2. \tag{5.16}$$

These choices are convenient for (p,p′) and (d,p) reactions, respectively (see Sect. 5.2). The cross sections are computed in units of fm^2/steradian, where 1 fm^2 = 10^{-26} cm^2.

The physical cross sections, which include reaction strengths and sums over the spin indices, are

$$\frac{\mathrm{d}\sigma^{lsj}(\theta)}{\mathrm{d}\Omega} = \frac{2J_B+1}{2J_A+1} \frac{2l+1}{2j+1} |B_{lsj}|^2 \sigma^{lsj}_{DW}(\theta) \qquad \text{for equal masses} \tag{5.17}$$

and

$$\frac{\mathrm{d}\sigma^{lsj}(\theta)}{\mathrm{d}\Omega} = \frac{2J_B+1}{2J_A+1} \frac{1}{2j+1} \frac{|B_{lsj}|^2}{10^4} \sigma^{lsj}_{DW}(\theta) \qquad \text{for unequal masses.} \tag{5.18}$$

In addition to the cross sections, DWUCK4 calculates certain spin observables. If we define a general spin operator $\sigma_{m_a m_a';m_b m_b'}$, then the program will also calculate the expectation value of the spin operator by the following expression:

$$\langle\sigma\rangle = \frac{\sum_{mm_am_b}\sum_{m_a'm_b'} S_{lsj}^{mm_am_b*}(\sigma)_{m_am_b;m_a'm_b'} S_{lsj}^{mm_a'm_b'}}{\sum_{mm_am_b}|S_{lsj}^{mm_am_b}|^2}. \tag{5.19}$$

The spin operator for the polarization of the final-state projectile is

$$(\sigma_y)_{m_am_a';m_bm_b'} = (\sigma_y)_{m_bm_b'}\delta_{m_a,m_a'} \qquad \text{for spin-}\tfrac{1}{2}$$

and

$$(\sigma_y)_{m_am_a';m_bm_b'} = (S_y)_{m_bm_b'}\delta_{m_a,m_a'} \qquad \text{for spin-1.}$$

Similarly, the spin operator for the polarization of the initial state projectile is

$$(\sigma_y)_{m_am_a';m_bm_b'} = (\sigma_y)_{m_am_a'}\delta_{m_b,m_b'} \qquad \text{for spin-}\tfrac{1}{2}$$

and

$$(\sigma_y)_{m_am_a';m_bm_b'} = (S_y)_{m_am_a'}\delta_{m_b,m_b'} \qquad \text{for spin-1.}$$

In the case of the initial state, the polarization is also the asymmetry observable.

For the case of spin-1 particles, tensor analyzing powers are also calculated. The operators for these quantities are defined elsewhere [5.4] and will not be given here.

5.2 Specific Cases of Reactions

5.2.1 Inelastic Excitation Using a Collective Nuclear Model

To first order in the deformation β_l, the interaction V can be written as

$$V(r,\theta) = -\beta_l \frac{R_0}{a}\frac{\mathrm{d}V(x)}{\mathrm{d}x} Y_l^0(\theta), \tag{5.20}$$

where $x = (r - R_0)/a$, $V(x)$ is the projectile–nucleus potential, and θ is the angle between the radius vector of the projectile and the nuclear symmetry axis. If we take the strong coupling form for the initial and final wave functions for the nucleus [5.3],

$$\psi_{JMK}^j = \sqrt{\frac{2J+1}{16\pi^2}}[\phi_K^j D_{K,M}^{J*}(\alpha,\beta,\gamma) + (-1)^{J-j}\phi_{-K}^j D_{-K,M}^{J*}(\alpha,\beta,\gamma)], \tag{5.21}$$

we can write

$$\langle J_B M_B s_b m_b | V | J_A M_A s_a m_a \rangle = -\beta_l \frac{R_0}{a} \frac{\mathrm{d}V(x)}{\mathrm{d}x} \sqrt{\frac{2J_A+1}{2J_B+1}}$$
$$\times (J_A\, l\, K\, 0 | J_B\, K)(J_A\, l\, M_A\, M_B - M_A | J_B\, M_B) Y_l^{m*}(\hat{\mathbf{r}}). \quad (5.22)$$

This form of the interaction gives the strength as

$$B_{l0l} = \beta_l \sqrt{\frac{2J_A+1}{2J_B+1}} (J_A\, l\, K\, 0 | J_B\, K), \quad (5.23)$$

and the resulting cross section is

$$\frac{\mathrm{d}\sigma^{lsj}(\theta)}{\mathrm{d}\Omega} = \beta_l^2 (J_A\, l\, K\, 0 | J_B\, K)^2 \sigma_{DW}^{l0l}(\theta). \quad (5.24)$$

In many cases, the inelastic scattering cross section is strongly affected by the Coulomb portion of the interaction between the projectile and target nucleus. This effect may be incorporated by a non-zero value of the parameter COUEX in input block 4 (see Sect. 5.3 below). An additional contribution is added to the form factor during computation of the radial integrals of the form

$$\begin{aligned} f_{l0l}^{\text{coulomb}}(r) &= \text{COUEX} \frac{3 * \text{Z} * \text{ZT} * \text{e}^2}{2l+1} \frac{R_c^l}{r^{l+1}} \quad \text{for } r > R_c \\ &= 0 \quad \text{for } r < R_c\,, \end{aligned} \quad (5.25)$$

where the parameters Z, ZT, and $R_c = r_{0c} * \text{MT}^{\frac{1}{3}}$ are the charges and charge radius from the kinematic input line in input block 7. In this way, the radius of the charge distribution for the form factor can be different from the charge radius used for the distorted waves. A value of COUEX = 1.0 will normalize the Coulomb amplitude in the conventional manner [5.1].

5.2.2 The (d,p) Stripping Reaction

The matrix element for a stripping reaction $A(a,b)B$ where $a = b + x$ and $B = A + x$ may be written as

$$\begin{aligned} &\langle J_B M_B s_b m_b | V | J_A M_A s_a m_a \rangle \\ &= \textstyle\sum_{jl} S_{jl}^{\frac{1}{2}} R_{jl}(r_{xA}) (l\, s\, m\, \mu - m | j\, \mu)(s_b\, s\, m_b\, m_a - m_b | s_a\, m_a) \\ &\times (J_A\, j\, M_A\, M_B - M_A | J_B\, M_B) D(\mathbf{r}_{\mathbf{xb}}) Y_l^{m*}(\hat{\mathbf{r}}_{\mathbf{xA}}), \end{aligned} \quad (5.26)$$

where $S_{jl}^{\frac{1}{2}}$ is the spectroscopic amplitude, $R_{jl}(r_{xA})$ is the radial wave function for the transferred particle x in the target nucleus B, and $D(\mathbf{r}_{\mathbf{xb}})$ is the product of the projectile internal function and the interaction potential between the projectile components x and b. In order to evaluate the DWBA matrix element, one makes use of the zero-range approximation,

$$D(\boldsymbol{r}_{\mathbf{xb}}) = D_0 \delta(\boldsymbol{r}_{\mathbf{x}} - \boldsymbol{r}_{\mathbf{b}}). \quad (5.27)$$

The reaction strength factor B_{lsj} becomes for this case

$$B_{lsj} = S_{jl}^{\frac{1}{2}} D_0, \tag{5.28}$$

where $S_{jl}^{\frac{1}{2}}$ is the spectroscopic amplitude. If the radial form factor in DWUCK4 is

$$f_{lsj} = R_{jl}(r_{xA}), \tag{5.29}$$

then the resulting cross section is written as

$$\frac{\mathrm{d}\sigma^{lsj}(\theta)}{\mathrm{d}\Omega} = \frac{2J_B+1}{2J_A+1}\frac{S_{lsj}}{2j+1}\frac{D_0^2}{10^4}\sigma_{DW}^{lsj}(\theta). \tag{5.30}$$

Typical deuteron models [5.1] give a value $D_0^2 \approx 1.5 \times 10^4$ MeV · fm^3.

A first-order correction to the zero-range approximation for single-particle transfer reactions may be made by multiplying the form factor f_{lsj} by the function

$$\begin{aligned} W_{\mathrm{FR}}(r) &= [1+A(r)]^{-1} \qquad \text{(Hulthén form)} \\ &= \exp[-A(r)] \qquad \text{(Gaussian form)}, \end{aligned} \tag{5.31}$$

where

$$A(r) = \frac{2}{\hbar^2}\frac{m_b m_x}{m_a} R^2[E_b - V_b(r_b) + E_x - V_x(r_x) - E_a + V_a(r_a)]. \tag{5.32}$$

Here R is the finite-range parameter, and E_a, E_b, E_x and V_a, V_b, V_x are the energies and potentials with respect to the target core for the light particles a, b and x. A positive value of the finite-range parameter FNRNG in input block 4 will select the Hulthén form, while a negative value will select the Gaussian form. Again typical models for the (d,p) reaction [5.1] give R ≈ 0.70 fm.

A non-local correction factor may also be needed by the use of a local equivalent potential for the distorted waves and for the transferred particle bound state. This factor again multiplies the form factor and for each particle is given by

$$W_{\mathrm{NL}}(r_i) = \exp\left[\frac{\beta_i^2}{8}\frac{2m_i}{\hbar^2}V_i(r_i)\right], \tag{5.33}$$

where β_i is the non-locality parameter for particle i and is input on the kinematics input line as PNLOC in input blocks 5, 6, and 7.

5.2.3 The (p,d) Pickup Reaction

The pickup reaction is the inverse of the stripping reaction described in the preceding section. Hence, we may write the cross section for the (p,d) reaction in terms of the (d,p) reaction cross section, using detailed balance, as

$$\begin{aligned}\frac{d\sigma_{pd}^{lsj}(\theta)}{d\Omega} &= \frac{2J_A+1}{2J_B+1}\frac{2s_a+1}{2s_b+1}\frac{k_a^2}{k_b^2}\frac{d\sigma_{dp}^{lsj}(\theta)}{d\Omega}\\ &= \frac{3}{2}\frac{S_{lsj}}{2j+1}\frac{D_0^2}{10^4}\sigma_{DW}^{lsj}(\theta),\end{aligned}\tag{5.34}$$

where σ_{DW}^{lsj} is the output from DWUCK4 with protons in the incident channel and deuterons in the exit channel.

5.2.4 The Microscopic Interaction Model for Inelastic Scattering

Central Interactions

This model considers the potential between the projectile and one of the nucleons in the target nucleus. It is convenient to expand this interaction in a Legendre polynomial series,

$$\begin{aligned}V(\boldsymbol{r}-\boldsymbol{r}_1) &= V_0\sum_l(2l+1)v_l(r,r_1)P_l(\hat{\mathbf{r}}\cdot\hat{\mathbf{r}}_1)\\ &= 4\pi V_0\sum_l v_l(r,r_1)\sum_m Y_l^m(\hat{\mathbf{r}})Y_l^{m*}(\hat{\mathbf{r}}_1).\end{aligned}\tag{5.35}$$

Using this form of the interaction, we find for the matrix element of V

$$\begin{aligned}&\langle J_BM_Bs_bm_b|V(\boldsymbol{r}-\boldsymbol{r}_1)|J_AM_As_am_a\rangle =\\ &\quad 4\pi V_0\textstyle\sum_l(J_A\,l\,M_A\,M_B-M_A|J_B\,M_B)\langle J_B||v_l(r,r_1)\mathrm{i}^lY_l(\hat{\mathbf{r}}_1)||J_A\rangle\\ &\quad\times(s_b\,s\,m_b\,m_a-m_b|s_a\,m_a)(-\mathrm{i})^lY_l^m(\hat{\mathbf{r}})^*\frac{1}{\sqrt{2J_B+1}}.\end{aligned}\tag{5.36}$$

The program DWUCK4 calculates the following expression for the radial factor [see (5.8)]:

$$\begin{aligned}f_l(r) &= 4\pi V_0\sqrt{2J_A+1}\langle j_1'j_2J_B||Y_l(\hat{\mathbf{r}}_1)||j_1j_2J_A\rangle\\ &\quad\times\int R_{l_1'j_1'}(r_1)v_l(r,r_0)R_{l_1j_1}(r_1)r_1^2\mathrm{d}r_1,\end{aligned}\tag{5.37}$$

where the $R_{lj}(r_1)$ are normalized radial functions for the initial and final states of the nucleon in the target with angular momentum l and total angular momentum j, and $\langle j_1'j_2J_B||Y_l(\hat{\mathbf{r}}_1)||j_1j_2J_A\rangle$ is the reduced matrix element as defined by Edmonds [5.2]. The angular momentum j_2 is the coupled angular momentum of the $A-1$ spectator nucleons. The program prints out (RME in the output) the quantity $\sqrt{4\pi}\langle j_1'j_2J_B||Y_l(\hat{\mathbf{r}}_1)||j_1j_2J_A\rangle$.

If there is more than one particle in each configuration then the form factor must be scaled by the amplitude of that configuration and by the counting factor, which takes into account the number of identical particles in the configuration. Thus, the strength of the reaction B_{l0l} is

$$B_{l0l}(j_1' j_2 J_B; j_1 j_2 J_A) = a_{j_1' j_2 J_B} a_{j_1 j_2 J_B} \times \sum_{i=1}^{n} \langle j_1' j_2 J_B || Y_l(\hat{\mathbf{r}}_i) || j_1 j_2 J_A \rangle \langle j_1' j_2 J_B || Y_l(\hat{\mathbf{r}}_1) || j_1 j_2 J_A \rangle. \tag{5.38}$$

Here, $a_{jj'J}$ are the amplitudes of the initial and final configurations that involve the factor for the n identical particles in the shell. If the strength of each configuration is scaled by B_{l0l}, then the cross section is

$$\frac{d\sigma_{pd}^{lsj}(\theta)}{d\Omega} = \sigma_{DW}^{l0l}(\theta). \tag{5.39}$$

Spin Dependent $\sigma_i \cdot \sigma_j$ Interaction

The interaction for this situation is given by the following form

$$V(\boldsymbol{r} - \boldsymbol{r}_1)\sigma_i \cdot \sigma_j = 4\pi V_0 \sum_{lj} v_l(r, r_1) \sum_{\mu} \mathcal{Y}_{l1j}^{-\mu}(\hat{\mathbf{r}}) \mathcal{Y}_{l1j}^{\mu *}(\hat{\mathbf{r}}_1), \tag{5.40}$$

where

$$\mathcal{Y}_{l1j}^{\mu}(\hat{\mathbf{r}}) = \sum_m (l\, s\, m\, \mu - m | j\, \mu) \mathrm{i}^l Y_l^m(\theta, \phi) \sigma_s^{\mu - m}, \tag{5.41}$$

and $\sigma_s^{m'}$ is the spin operator for spin s and projection m'. The matrix element for the transition amplitude then becomes

$$\begin{aligned} &\langle J_B M_B s_b m_b | V(\mathbf{r} - \mathbf{r}_1) | J_A M_A s_a m_a \rangle \\ &= 4\pi V_0 \textstyle\sum_{lj\mu} (J_A\, j\, M_A\, M_B - M_A | J_B\, M_B) \langle J_B || v_l(r, r_1) \mathrm{i}^l \mathcal{Y}_{l1j}(\hat{\mathbf{r}}_1) || J_A \rangle \\ &\times (-1)^{2j} 2\sqrt{s_s(s_a + 1)} (l\, s\, m\, \mu - m | s_a\, m_a) \\ &\times (s_b\, s\, m_b\, m_a - m_b | s_a\, m_a)(-\mathrm{i})^l Y_l^m(\hat{\mathbf{r}})^* \frac{1}{\sqrt{2J_B + 1}}, \end{aligned} \tag{5.42}$$

where for the usual microscopic interaction, $s = 1$ and $s_a = s_b$.

The program DWUCK4 computes the form factor

$$\begin{aligned} f_{lsj}(r) &= 2\sqrt{s_a(s_a + 1)} \mathrm{i}^l \sqrt{\frac{2l + 1}{(2j + 1)(2J_A + 1)}} \\ &\times \langle j_1' j_2' J_B || v_l(r, r_1) \mathcal{Y}_{l1j}(\hat{\mathbf{r}}_1) || j_1 j_2 J_A \rangle \end{aligned} \tag{5.43}$$

In the above expression, we have used the reduced matrix element [5.2]

$$\langle s_a || \sigma || s_a \rangle = 2\sqrt{s_a(s_a + 1)}.$$

If the spin of the projectile changes such as in the (^{6}Li,^{6}He) reaction, then the matrix elements must be scaled by the factor

$$\frac{\langle s_b || \sigma || s_a \rangle}{2\sqrt{s_a(s_a + 1)}}.$$

5.3 Description of the Input for DWUCK4

The input to the program is defined by seven input blocks, several of which require more than one line of input. The first four input blocks specify the basic input to the program. The next two blocks, 5 and 6, are multiline blocks that specify the initial and final distorted waves and are structured so that potentials consisting of several terms may be "stacked" and accumulated. Finally, the seventh block describes the form factor and its input depends on the particular reaction model chosen. The input scheme has been designed for flexibility, hence some redundancy in the input is present and input consistency is not enforced. The formats are presented in their FORTRAN 77 form. In many computer-system applications, an abbreviated form with comma separators is allowed, for example, a valid form for input block 2 is 36. ,0. ,5.

Input (1 line) Block 1 ICON(20), ALPHA FORMAT(20I1, A60)

i	ICON(i)	Description
1	0	Do not read in input block 2 (use default or the angle data of the previous case).
	1	Read input block 2 (angle data).
	9	Stop program and exit.
2	0	Collective or particle-transfer nuclear model.
	2	Microscopic inelastic nuclear model.
3	0	Use the same radial form factor for each l-transfer (read in only one set of input block 7).
	1	Compute a separate radial form factor for each l-transfer (read in one input block 7 for each l-transfer).
	2	Same as ICON(3) = 1 except that the cross section is the coherent sum of the amplitudes from each l-transfer.
4	0	Output each radial form factor before finite-range and non-local corrections.
	1	Suppress output of the radial form factor.
	2	Suppress output of radial form factor and intermediate output in two-particle form factors.
	3	Output form factor after calculation of the finite range and non-local correction factors.
5	0	Suppress output of the elastic scattering amplitudes for the distorted waves.
	1	Output the elastic scattering amplitudes $[\exp(2i\delta_{lj})-1]/2i$.

6	0	Suppress output of the elastic scattering cross sections for the distorted waves.
	1	Output the elastic scattering cross sections for the distorted waves.
7	0	Suppress output of the radial matrix elements.
	1	Output the radial matrix elements as defined in (5.13).
8	0	Suppress output of the inelastic scattering amplitudes.
	1	Output the inelastic scattering amplitudes as defined by (5.12).
9	0	Suppress output plotting of the inelastic cross section.
	N	Output a paper plot of the inelastic cross section as an N-decade logarithmic plot.
10	0	The kinematics for the waves are calculated non-relativistically.
	1	The kinematics for the waves are calculated relativistically.
11		Not used
12	0	Suppress output of the radial wave functions for the distorted waves.
	N	Output the radial wave functions for the distorted waves at every Nth radial point.
13	0	Suppress output of the inelastic cross section on a separate file.
	1	Output the inelastic cross section and angles to the file `FOR007.DAT`.
14		Not used.
15	0	Suppress output of $K(r)^2$, the square of the local wave number for each distorted wave.
	1	Output $K(r)^2$, the square of the local wave number for each distorted wave.
16	0	Suppress output plotting of the elastic scattering cross sections for the distorted waves.
	N	Output a paper plot of elastic scattering cross sections for the distorted waves as an N-decade logarithmic plot.
Alpha		Any 60 characters (including blanks) to identify the run, beginning in column 21.

Input Block 2 (1 line)	NANGLES, ANGLE1, DANGLE FORMAT(3F8.4)
NANGLES	Number of angles for calculating the cross sections.
ANGLE1	First angle (degrees). This may be zero.
DANGLE	Angle increment (degrees).

The program has a default set of angle data for the interval of 0° to 180° in 5° intervals that is used if no angle data are read in.

Input Block 3	LMAX, NLTR, [LTR(I), I=1,NLTR], [JTR(I), I=1,NLTR]	FORMAT(18I3)
(1 line)		
LMAX	Maximum partial wave for the distorted waves. The limit imposed by the storage is $400/(2s_a + 2s_b + 2) - 1$, where s_a and s_b are the spins of the initial and final projectiles.	
NLTR	The maximum number of angular-momentum transfer to be calculated is NLTR ≤ 8.	
LTR(I)	The angular transfer for the Ith form factor.	
JTR(I)	Twice the total angular momentum transfer for the Ith radial form factor.	

Input Block 4	DR, RMIN, RMAX, COUEX, FNRNG	FORMAT(10F8.4)
(1 line)		
DR	Integration step size for the radial coordinate (in fm).	
RMIN	Lower cutoff radius for the radial integrals (in fm).	
RMAX	Upper cutoff for the radial integrals (in fm). If RMAX is negative, the automatic resizing of RMAX in the program is overridden and the input value of |RMAX| is used. The storage allocation in the program restricts the number of radial points to INT(RMAX/DR) ≤ 400.	
COUEX	Coulomb excitation scale factor.	
FNRNG	Finite-range correction factor.	

Input Block 5	Initial distorted wave input block	
(minimum of 2 lines)		
Input line 1	E, MP, ZP, MT, ZT, r_{0c}, AC, PNLOC, 2*FS	FORMAT(10F8.4)
(Kinematic input line)		
E	Laboratory energy (in MeV) of the initial projectile (must be > 0).	
MP	Projectile mass (in AMU units).	
ZP	Projectile charge.	
MT	Target mass (in AMU units).	
ZT	Target charge.	
r_{0c}	Reduced charge radius ($R_c = r_{0c}\, A^{\frac{1}{3}}$) in fm.	

AC	Diffuseness of charge radius (not implemented).
PNLOC	Nonlocality parameter.
2*FS	Twice spin of the projectile.
Input lines (2,...,)	OPTION, VR, r_{0R}, AR, VSOR, VI, r_{0I}, AI, VSOI, POWER FORMAT(10F8.4)
(Potential input)	
OPTION	Option number for potential shape (these options are defined below).
VR	Strength of the real potential (in MeV).
r_{0R}	Reduced radius of the real potential ($R_R = r_{0R}\ A^{\frac{1}{3}}$) (in fm).
AR	Diffuseness of the real potential (in fm).
VSOR	Spin–orbit factor for the real potential.
VI	Strength of the imaginary potential (in MeV).
r_{0I}	Reduced radius of the imaginary potential ($R_I = r_{0I}\ A^{\frac{1}{3}}$) (in fm).
AI	Diffuseness of the imaginary potential (in fm).
VSOI	Spin–orbit factor for the imaginary potential.
POWER	Extra parameter used in some potential and form-factor options.

Any number of potential options may be used on successive lines and the resulting potential will be the superposition of the individual potentials. If OPTION is a negative number, then that option will be the last potential. The program will then continue to the next input block. If the OPTION is zero, then no potential will be computed and the program will process the next input block.

Input Block 6	Final distorted-wave input block
(minimum of 2 lines)	
Input line 1	QCODE, MP, ZP, MT, ZT, r_{0c}, AC, PNLOC, 2*FS FORMAT(10F8.4)
(kinematic input line)	
QCODE	Q-value for the reaction (in MeV).

The remainder of the parameters on this line are interpreted in exactly the same manner as for input block 5 but describe the kinematics and potential for the final distorted wave. The kinematic input line is then followed by the lines (2, ...,) describing the potential as in input block 5.

Input Block 7	The radial form factor for the collective model and particle-transfer reactions.
(minimum of 2 lines)	

Input line 1	E, MP, ZP, MT, ZT, r_{0c}, AC, PNLOC, 2*FS FORMAT(10F8.4)
(Kinematic input line)	
E	Binding energy of the single particle (in MeV).
MP	Single particle mass (in AMU units).
ZP	Single particle charge.
MT	Core nucleus mass (in AMU units).
ZT	Core nucleus charge.
r_{0c}	Reduced charge radius ($R_c = r_{0c}\ A^{\frac{1}{3}}$) (in fm).
AC	Diffuseness of charge radius (not implemented).
PNLOC	Non-locality parameter.
2*FS	Twice the spin transfer of the form factor.
Input lines(2, ...,)	OPTION, VR, r_{0R}, AR, VSOR, VI, r_{0I}, AI, VSOI, POWER FORMAT(10F8.4)
(Potential lines)	

These input lines have the same meaning as the same lines of input block 5.

If the parameter E in line 1 is zero, then the potential defined by lines (2, ...,) is taken to be the radial form factor. This is the standard option for the inelastic-scattering collective model. If the parameter E in the first line of this block is non-zero, then the program computes an eigenfunction solution using the potential defined by lines (2, ...,). In this case, the additional line of input below must be added to define the quantum numbers and other needed parameters.

Extra input line	FNODE, FL, 2*FJ, 2*FS, VTRIAL, FISW FORMAT(10F8.4)
(Kinematic input line)	
FN	Number of nodes in the radial function excluding the origin and infinity.
FL	Orbital angular momentum for the radial function.
2*FJ	Twice the total angular-momentum quantum number for the radial function.
2*FS	Twice the intrinsic spin of the radial function.
VTRIAL	Scaling factor for the potential defining the radial function. If VTRIAL is zero, a default value of +60.0 is used.
FISW	Search control for the integration of the radial function. = 0. Search on the strength of the potential scale factor VTRIAL keeping the binding energy E fixed. = 1. Search on the binding energy E keeping the potential strength VTRIAL fixed.

FISW (contd.) = 2. No search (valid for E > 0 only).
The radial wave function is calculated for the given energy E and strength VTRIAL and matched to the asymptotic solution with the calculated phase shift.

Since the total potential is a product of VTRIAL and the forms calculated by lines (2, ...,), the result must be negative in order to give the usual bound state function, that VTRIAL*VR must be less than zero (for OPT = 1.0).

Input Block 7 Description of the radial form factor for the microscopic inelastic scattering model.
[Used when ICON(2) = 2]
Input line 1 CONTROL, OPCODE, FLMU, VZERO, FJ2, FJI, FJF FORMAT(10F8.4)
(Option input line)

CONTROL = 0.0, read no input blocks for the single particle and terminate the form factor calculation.
= 1.0, read one input block 7 for single particles and use this radial function for both particles.
= 2.0, read two input blocks 7, one for each single particle.

OPCODE = 0.0, not an option.
= 1.0, calculate form factor using a Yukawa potential $V(r) = V_0 \exp(-\mu r)/(\mu r)$.
= 2.0, calculate form factor using a Coulomb potential.
= 3.0, calculate form factor using a tensor-force potential.
= 4.0, not used.
= 5.0, calculate a microscopic two-nucleon transfer form factor.
= 6.0, calculate a zero-range knockout form factor.

FLMU = Inverse range parameter (μ) in the inelastic scattering model or rms radius of a Gaussian wave function used in two-nucleon-transfer form factor.

VZERO = strength of the potential used in options 1, 2, and 3 above.
= amplitude of the configuration for the two-nucleon-transfer reaction of option 5.
= volume integral of the two-body potential in the zero-range knockout model in option 6.

FJ2 = Twice spin of the core (j_2) to which the active single particle is coupled.

FJI	= Twice the spin of the initial nucleus $\lvert \boldsymbol{j}_1 + \boldsymbol{j}_2 \rvert$ = JI (J_A in Sect. 5.2.1)
FJF	= Twice the spin of the initial nucleus $\lvert \boldsymbol{j}'_1 + \boldsymbol{j}_2 \rvert$ = JF (J_B in Sect. 5.2.1)

The values for the spins j_1 and j'_1 are taken from the input line defining the quantum numbers in the input block 7 specified below.

Input lines (2, ...,) Following input line 1 insert one or two input blocks (depending on the value of CONTROL in line 1) defining the single-particle orbitals. The input lines are identical to those of input block 7 for the single-particle-transfer case. The program will add the form factors coherently until it encounters a negative or zero value for CONTROL.

Potential options available for input blocks 5, 6, and 7.

OPT = 1.0 Volume Wood–Saxon potential

$$V(r) = V_{\mathrm{R}} f(x_{\mathrm{R}}) + \mathrm{i} V_{\mathrm{I}} f(x_{\mathrm{I}}),$$

where

$$f(x_i) = 1.0/[1 + \exp(x_i)];\ x_i = (r - r_{0i}\mathrm{MT}^{\frac{1}{3}})/A_i.$$

Note that in order to have an attractive real and absorptive imaginary potential, both VR and VI must be negative.

OPT = 2.0 Surface Wood–Saxon (or derivative) potential.

$$V(r) = V_{\mathrm{R}} g(x_{\mathrm{R}}) + \mathrm{i} V_{\mathrm{I}} g(x_{\mathrm{I}});\ g(x) = \frac{\mathrm{d}f(x)}{\mathrm{d}x},$$

where x is defined the same as for OPT = 1.0. This form of the potential has *no* factors of 4 in its definition so that the strength V_{I} is related to W_D by $V_{\mathrm{I}} = 4W_D$. Further, in order that the potential have attractive real and absorptive imaginary properties, V_{R} and V_{I} must be positive.

OPT = 3.0 Second-derivative Wood–Saxon potential.

$$V(r) = V_{\mathrm{R}} h(x_{\mathrm{R}}) + \mathrm{i} V_{\mathrm{I}} h(x_{\mathrm{I}});\ \ h(x) = \frac{\mathrm{d}^2 f(x)}{\mathrm{d}x^2}.$$

OPT = 4.0 Spin–orbit potential from a volume Wood–Saxon form

$$V_{LS}(r) = \left[-V_{\mathrm{R}} \frac{1}{r} \frac{\mathrm{d}f(x_{\mathrm{R}})}{\mathrm{d}r} - \mathrm{i}V_{\mathrm{I}} \frac{1}{r} \frac{\mathrm{d}f(x_{\mathrm{I}})}{\mathrm{d}r} \right] \mathbf{L} \cdot \mathbf{s}.$$

The above potential is defined in terms of $\mathbf{L} \cdot \mathbf{s}$ without any $(\hbar/m_\pi c)^2 \sim 2$ factor. The strength V_{R} is about four times the strength of the forms defined with the $(\hbar/m_\pi c)^2$ factor and an $\mathbf{L} \cdot \sigma$ operator. The VSOR and VSOI parameters in the input line for the OPT = 1.0 case will also give a spin–orbit potential using the same geometry. The VSOR and VSOI are interpreted as a non-dimensional multiplier (λ in the so-called Thomas form [5.1]). The potential is written as

$$V_{LS}(r) = \left[-V_{\mathrm{R}} \frac{\mathrm{VSOR}}{45.2} \frac{\mathrm{d}f(x_{\mathrm{R}})}{r\mathrm{d}r} - \mathrm{i}V_{\mathrm{I}} \frac{\mathrm{VSOI}}{45.2} \frac{\mathrm{d}f(x_{\mathrm{I}})}{r\mathrm{d}r} \right] \mathbf{L} \cdot \mathbf{s}.$$

OPT = 5.0 Spin–orbit potential from a surface Wood–Saxon form

$$V_{LS}(r) = \left[-V_{\mathrm{R}} \frac{1}{r} \frac{\mathrm{d}g(x_{\mathrm{R}})}{\mathrm{d}r} - \mathrm{i}V_{\mathrm{I}} \frac{1}{r} \frac{\mathrm{d}g(x_{\mathrm{I}})}{\mathrm{d}r} \right] \mathbf{L} \cdot \mathbf{s}.$$

OPT = 6.0 Volume Wood–Saxon potential with an r^{POWER} factor

$$V(r) = [V_{\mathrm{R}} f(x_{\mathrm{R}}) + \mathrm{i}V_{\mathrm{I}} f(x_{\mathrm{I}})] r^{\mathrm{POWER}}.$$

OPT = 7.0 Surface Wood–Saxon potential with an r^{POWER} factor

$$V(r) = [V_{\mathrm{R}} f(x_{\mathrm{R}}) + \mathrm{i}V_{\mathrm{I}} f(x_{\mathrm{I}})] r^{\mathrm{POWER}}.$$

OPT = 8.0 External potential or form factor: A potential or form factor will be read in from the input file. In addition to the potential option line, an additional line will specify the number of radial points and whether the function is to be added to the real or imaginary part of the potential or form factor. Then the lines containing the function will be read in. The radial increment of the function that is read in must match that of the distorted wave or form factor. In addition, a real function will be scaled by VR and an imaginary function will be scaled by VI. If the VR or VI value is zero or left blank, the function will not be scaled.

Input line 1	F1, F2	FORMAT(10F8.4)
F1	Number of radial points to be read in.	
F2	= 0.0 considers the function to be real.	
	= 1.0 considers the function to be imaginary.	
input lines (2, ...,)	[FF(I), I=1,F1]	FORMAT(5E16.7)

OPT = 9.0 Normalized harmonic oscillator:

$$V(r) = V_{\mathrm{R}} N \mathcal{L}(r/r_{\mathrm{0R}}) \exp[-\frac{1}{2}(r/r_{\mathrm{0R}})^2],$$

where $\mathcal{L}(x)$ is a Laguerre polynomial and N is the normalization constant for the functions such that

$$N^2 \int_0^\infty [\mathcal{L}(r/r_{\mathrm{0R}})]^2 \exp[-(r/r_{\mathrm{0R}})^2] r^2 \mathrm{d}r = 1.$$

Note that the radius parameter is the reduced radius r_{0R}.

OPT = 10.0 Gaussian r^{POWER}.

$$V(r) = V_{\mathrm{R}} \exp[-(r/r_{\mathrm{0R}})^2] r^{\mathrm{POWER}}.$$

Again note that the radius parameter is the reduced radius r_{0R}.

OPT = 11.0 Legendre polynomial expansion of a volume Wood-Saxon potential:

$$V(r) = \int \{V_{\mathrm{R}} f[x_{\mathrm{R}}(r,\theta)] + \mathrm{i} V_{\mathrm{I}} f[x_{\mathrm{I}}(r,\theta)]\} Y^0_{\mathrm{LTR}}(\theta) \mathrm{d}\Omega_r,$$

where

$$f(x) = \frac{1}{1+\exp(x)}$$

and

$$x_i = [r - R_i(1+\beta_\lambda Y^0_\lambda)]/a_i.$$

Here LTR is the orbital angular transfer for this form factor, λ, the value of LAM, is the order of the multipole of the deformation and β_λ is the deformation parameter. An extra input line must follow this potential option with the values of β_λ and LAM.

Extra input line	BETA, LAM	FORMAT(2F8.4)
BETA	Value of β_λ, the deformation parameter.	
LAM	Value of the deformation order λ which can be different from LTR.	

OPT = 12.0 Legendre polynomial expansion of a surface Wood–Saxon potential:

$$V(r) = \int \left\{ V_{\mathrm{R}} \frac{\mathrm{d}f[x_{\mathrm{R}}(r,\theta)]}{\mathrm{d}x_{\mathrm{R}}} + \mathrm{i}V_{\mathrm{I}} \frac{\mathrm{d}f[x_{\mathrm{I}}(r,\theta)]}{\mathrm{d}x_{\mathrm{I}}} \right\} Y^{0}_{\mathrm{LTR}}(\theta)\mathrm{d}\Omega_r,$$

where $f(x)$ and x have the same meanings as in option 11. In both of the options 11 and 12, the potentials for the distorted waves are calculated for LTR = 0 only, but LAM can have any value, but when these options are used as form factors the angular momentum transfer LTR is used and can take on any value.

Form factor options available for input block 7

These options are generally used for the collective model cases, that is, when ICON(2) = 0 and when the energy parameter E = 0.0 in the first line of input block 7. These options have the same basic form as in input blocks 5 and 6 above except for the following modifications that have been made for convenience in treating the collective model.

OPT = 1.0 Volume Wood–Saxon potential

$$\begin{aligned} V(r) &= V_{\mathrm{R}} f(x_{\mathrm{R}})(R_{\mathrm{R}}/a_{\mathrm{R}})^{(\mathrm{POWER}+1.0)} \\ &\quad + \mathrm{i}V_{\mathrm{I}} f(x_{\mathrm{I}})(R_{\mathrm{I}}/a_{\mathrm{I}})^{(\mathrm{POWER}+1.0)}. \end{aligned}$$

The default value for POWER = 0.0 gives the usual factor of (R/a) in the definition of the collective model.

OPT = 2.0,3.0 As in the case for OPT = 1.0, the real and imaginary components of the form factor are multiplied by their appropriate $(R/a)^{(\mathrm{POWER}+1.0)}$ factor.

OPT = 4.0,5.0 These options are not available as form factors.

OPT = 6.0-12.0 These options give the same functional form and normalizations as in the same numbered options in input blocks 5 and 6.

5.4 Sample Cases–Input and Output

Two cases of sample input are included on the disk (file `DWUCK4.INP`). The first example describes the excitation of a 3^- excited state of ^{56}Fe by the inelastic scattering of 22.5 MeV protons using a collective model where the form factor is the first derivative of the optical potential. Here the Coulomb excitation contribution is included. The second example gives the input for a ^{40}Ca(d,p)^{41}Ca reaction where the model assumes the stripping of a $1f_{7/2}$

Table 5.1. Results for the two example cases defined above

Value	^{56}Fe(p,p′)^{56}Fe	^{40}Ca(d,p)^{41}Ca
REACSIG 1	1.1276E+02	1.3394E+02
REACSIG 2	1.0931E+02	9.6707E+01
Tot-Sig	2.9046E+01	1.8393E+00

neutron. The calculation includes a finite-range correction with a radius parameter of 0.70 fm. In both cases, the optical potentials for the projectiles are composed of volume-real and surface-imaginary potentials and also include spin–orbit potentials. The final input line (beginning with a 9) signals the end of data.

The first page of the output echoes the input blocks 1–6 and displays the derived kinematical quantities for the distorted waves such as the center-of-mass energies, wave numbers, Coulomb parameters, etc. The second page outputs the same information for input block 7 including output of the radial form factor. The third page gives output for the inelastic cross sections and spin observables in the first six columns; the next six columns give the same output appropriate to the time-reversed reaction. The final two pages of output are paper plots of the cross section and asymmetry parameter, which is the polarization for the time-reversed reaction. The file `DWUCK4.OUT` contains a typical output, exemplified for the example cases discussed above.

Typical numbers to check in the output are the absorption cross section REACSIG for the distorted waves and the total inelastic cross section Tot-Sig. The correct value for REACSIG confirms that the elastic scattering has been calculated correctly. The correct value of Tot-Sig indicates that the reaction-theory part of the program has also been entered and calculated correctly. Values of the cross sections for these two cases are given in Table 5.1 in fm^2.

5.5 Things to Do

The user should first run the two sample cases above and verify that the output numbers check with those in the table. Following this it might be instructive to change the (d,p) input case to other values of transferred angular momenta (ℓ, j)– say $p_{3/2}$ and $p_{1/2}$– which will illustrate the strong angular-distribution dependence on ℓ and weak dependence on j. Similar runs can also be made for the (p,p′) case. Finally, for the analytically inclined, it is possible to obtain a closed-form solution for a plane-wave $\ell = 0$ case with a harmonic-oscillator form factor (option 9 above). This check can then be exploited by the user to understand in detail the normalization factors employed in the `DWUCK4` program.

5.6 Technical Note

The code is not intended to be run on a microcomputer. Therefore, it has not been converted to strict ANSI-standard FORTRAN 77. When running the code, it requires the assignment of an input and an output file. For use on a VAX computer, a convenient command file `DWUCK4.VAX` is supplied on the disk.

References

[5.1] G. R. Satchler, *Direct Nuclear Reactions* (Oxford University Press, Oxford, 1983)

[5.2] A. R. Edmonds, *Angular Momentum in Quantum Mechanics* (Princeton University, Princeton, 1957)

[5.3] Aage Bohr and Ben R. Mottelson, *Nuclear Structure* Vols. I and II (W. A. Benjamin, New York, 1969)

[5.4] H. H. Barschall and W. Haeberli, *Polarization Phenomena in Nuclear Reactions* (University of Wisconsin Press, Madison, 1969)

6. Statistical-Model Calculations with Angular-Momentum Coupling

A. Gavron

6.1 Background

The origins of the statistical model of the nucleus can be traced back to the seminal paper by Weisskopf, published in 1937 [6.1]. Weisskopf considers a nucleus A excited to an excitation energy E_A. The probability of emitting ("evaporating") a neutron with a kinetic energy ϵ, transforming the nucleus to B, with an excitation energy $E_B = E_A - B_n - \epsilon$, is calculated using the principle of detailed balance. This probability, which he denotes $W_n(\epsilon)$, can then be determined if the cross section for the inverse process, $\sigma(E_A, \epsilon)$, is known. This is the cross section for production of a compound nucleus, by collision of a neutron with nucleus B (with an excitation energy of $E_A - B_n - \epsilon$). In most calculations, one uses the cross sections obtained from ground-state compound-nucleus-formation measurements, or from inelastic cross sections obtained from optical-model calculations. The user should be aware that this is an approximation and that it has its limitations [6.2]. The expression for $W_n(\epsilon)$ is then

$$W_n(\epsilon) = \sigma(E_A, \epsilon) g m \epsilon \omega_B(E_B) / \pi^2 \hbar^3 \omega_A(E_A), \tag{6.1}$$

where m is the neutron mass, g is a spin degeneracy factor, and ω_A and ω_B are the level densities of the initial and final nuclei, respectively.

The Weisskopf expression has been successfully used in the many reactions where angular-momentum effects are not important. However, it is often necessary to consider the effects of angular momentum and parity conservation on the decay modes. One then needs to use the Hauser–Feshbach formalism [6.3] to calculate the decay probability of an excited nucleus with an excitation energy E and spin J into a specific exit channel. The probabilility is determined by the level density $\rho(\mathrm{E}_f, \mathrm{J}_f)$ of the residual nucleus at an excitation energy E_f and angular momentum J_f, and the transmission coefficient of the emitted particle $T_l(\epsilon)$, where l is the orbital angular momentum removed by the particle, I the particle spin, and ϵ is its kinetic energy. Energy and angular-momentum conservation pose the following constraints:

$$\begin{aligned} E &= E_f + \epsilon + BE_i, \\ \boldsymbol{J} &= \boldsymbol{I} + \boldsymbol{l} + \boldsymbol{J}_f, \end{aligned} \tag{6.2}$$

where BE_i is the binding energy of the emitted particle in decay channel i. The exact expression for the decay width Γ into a specific channel i is

$$\Gamma_i(E,J) = D(E,J)/(2\pi) \sum_{I,l,J_f} \delta(\boldsymbol{I}+\boldsymbol{l}+\boldsymbol{J}_f-\boldsymbol{J}) \times \int_0^{E-BE_i} T_l(\epsilon)\rho_i(E-BE_i-\epsilon, J_f)\mathrm{d}\epsilon \, , \tag{6.3}$$

where $D(E,J)$ is the level spacing (reciprocal of the level density) of the decaying nucleus at its initial excitation energy and spin. The relative decay probability into channel i is given by $\Gamma_i/\Gamma_{\text{tot}}$ where $\Gamma_{\text{tot}} = \sum_i \Gamma_i$, where i runs over all open channels.

A comprehensive understanding of the level-density dependence on the compound-nucleus angular momentum is essential for obtaining reliable results when significant angular momentum is present. An important step in this direction was taken by J. R. Grover [6.4], who coined the term "Yrast level" to describe the lowest-lying level at any given angular momentum: Since for any angular momentum J the nucleus has to have an excitation energy at least as high as the rotational energy $E_{\text{rot}}(J)$, there cannot be any states with an excitation energy below $E_{\text{rot}}(J)$. Thus, when a nucleus comes close to the "Yrast line," particle decay is inhibited (the level density is close to zero for the final states allowed by energy and angular-momentum conservation), and decay proceeds mainly by stretched gamma emission.

6.2 Basic Features of the Code

The statistical model code PACE2 (projection – angular-momentum-coupled-evaporation, version 2) we present here was derived from a predecessor, code JULIAN [6.5]. It uses a Monte Carlo procedure to determine the decay sequence of an excited nucleus, using the Hauser–Feshbach formalism. Many events ("cascades") are followed until the residual nucleus in each cascade reaches its ground state. At each stage, the decay probability is determined by calculating $2\pi\Gamma/D$ for neutron, proton, alpha, and gamma decay to final states differing in excitation energy, spin, nucleon number A, and proton number Z, as well as the fission probability using the Bohr–Wheeler [6.6] formalism. A random number selection determines the actual final state the nucleus decays to, and the process is then repeated for the other cascades, until all the nuclei reach the ground state.

A typical output from such a calculation would be the A and Z distribution of the residual nuclei, particle and photon spectra, fission probability, etc. More details can be obtained such as the particle spectra leading to a residual nucleus with a specific A and Z or particles preceding fission. A special feature of this code is its ability to provide angular distributions of evaporated particles and residues, obtained by tracking the distribution of the angular-momentum projection through each cascade. The code contains a summary routine STATIS, which the user can adapt to provide any specific output desired.

6.3 Level-Density Tables

The level-density table is one of the crucial ingredients of the calculation. There are two default level-density options that consist of a Fermi gas formalism that switches to a constant-temperature formalism at low energies: 1. The Gilbert and Cameron [6.7] procedure to determine the "little a" parameter (denoted a); 2. a is taken to be A/f where A is the number of nucleons and f is a constant factor. The rotational energy contribution, $E_{\text{rot}}(\text{J})$, can be selected independent of the option we choose for a: (i) the Gilbert and Cameron [6.7] spin-cutoff parameter can be used; (ii) the ground-state rotational energy of the finite-range rotating-drop model of A. J. Sierk [6.8] is available. For values of A, Z, or J beyond the range of validity of Sierk's routine, rotational energies are taken from the work of Cohen, Plasil, and Swiatecki [6.9]. The level density at a specific spin value J, $\rho(E, J)$ is given by

$$\rho(E, J) = (2J + 1)\omega[E - E_{\text{rot}}(J)], \tag{6.4}$$

where $\omega(E)$ is the total level density at excitation energy E. We note that this is different from the original Gilbert and Cameron formalism, which assumes that the spin dependence is factorizable. The formalism we use here is more accurate for high spins, close to the Yrast line.

We now discuss the structure of the level-density tables: This should facilitate the users' ability to write their own level-density routines for specific problems. The level density is stored in a vector RLEV. The spin and excitation energy at any position K in the vector are determined by the vectors EBIN, MAXJ, and MAXJS. The spin index j, which determines the memory location of spin J for a vector that is a function of angular momentum, is given by $J+1/2$ for odd nuclei and by $J+1$ for even nuclei. MAXJ(i) is the number of entries (due to different spins) to the level-density table at excitation energy EBIN(i). MAXJS(i) is the running sum of MAXJ(i), from 1 to i:

$$\text{MAXJS(i)} = \sum_{j=1}^{i} \text{MAXJ(j)}. \tag{6.5}$$

Thus, a level with excitation energy E [equal, for example, to EBIN(i)] and spin index j, will have a level density RLEV(K), where K=MAXJS(i)−MAXJ(i)+j. A look at the WRITE statements following line 2770 in the code may help clarify this point. As an additional example, an experimental ("discrete") level of energy 2.221 MeV and spin 3, in addition to the ground state (E=0, J=0), would necessitate the following entries:

EBIN(1)=0.000 MAXJ(1)=1 MAXJS(1)=1 RLEV(1)=1,
EBIN(2)=2.221 MAXJ(2)=4 MAXJS(2)=5 RLEV(2,3,4)=0. RLEV(5)=1.

Note the necessity of padding RLEV with zero values up to the necessary spin, and that spin 3 goes into position ("spin index") 4 for that excitation energy.

Calculating the level density at the saddle point (for fission probability calculations) involves an additional parameter: a_f/a_γ (ARATIO in the code). This parameter is used to determine a_f from the input value a (Gilbert and Cameron, or A/f).

When working with computers having a limited exponent range, it is necessary to renormalize the level density: The entire level-density table is multiplied by 10^{-25} to prevent exponent overflow.

6.4 Transmission Coefficients

The transmission coefficients are calculated at initialization time for the compound-nucleus values of A and Z. Optical-model potentials can be read in, or default values used. For protons and neutrons, default potentials were taken from Perey and Perey [6.10], and for alpha particles from Igo and Huizenga [6.11]. As the nucleus decays, we assume that the transmission coefficients (of charged particles) are shifted in their kinetic-energy dependence. We assume that the transmission coefficients are a function of ϵ - $V_c(A, Z)$, where A and Z are the nucleon and proton numbers of the emitting nucleus, and V_c the Coulomb barrier for the particle emitted.

Gamma emission is considered using E1, E2, M1, and M2 transition intensities. The code contains default values, but one should interpret the results with caution – only statistical gamma rays are emitted, whereas, in reality, a large part of the photon spectrum comes from rotational (collective) bands. The E1 and E2 values usually need to be adjusted even to obtain the correct *statistical* gamma ray multiplicity.

6.5 Fission Barrier and Probability

The fission probability is calculated using the Bohr–Wheeler saddle-point formalism [6.6]. Tunneling effects below the barrier (Hill–Wheeler formalism [6.12]) are not currently incorporated – consequently, the code should not be used for near-barrier calculations. The fission barriers are those of Sierk [6.8], and provision is made to multiply them by a constant (angular-momentum-independent) factor. Care has to be taken in regions of nuclei where strong shell effects influence the fission barrier, such as in the lead and actinide regions. It should still be possible to use the code for excitation energies above ~50 MeV where shell effects are expected to wash out. Alternatively, accurate barriers and level densities can be coded in to provide more accurate calculations.

A special feature of this program is its ability to calculate fission probabilities that are too low to actually produce fission events in the Monte Carlo sampling: This is done in subroutine STATIS by adding all the consecutive fission probabilities at each stage of the cascade.

6.6 Angular Distribution

The angular distribution of the emitted particles (and hence of the recoiling residual nuclei) is determined at each stage of deexcitation by the appropriate spherical harmonic $|Y_l^m(\theta,\phi)|^2$. l is the orbital angular momentum removed by the outgoing particle, determined by the transmission coefficients T_l for the decay. m is determined by the squares of the angular momentum coupling coefficients that couple (J_f, M_f) with (l, m), to form the initial spin and its projection, (J, M). In this calculation, we ignore the effect of the particle spin.

The code contains two different initial possibilities: 1. the initial nucleus is produced by a fusion reaction. The initial spin projection is $M = 0$ with respect to the quantization axis, which is taken to be the beam axis. θ is then the angle with respect to this axis. 2. the initial nucleus is produced aligned perpendicular to the reaction plane (as in a strongly damped collision). The initial spin projection is $M = J$ with respect to a quantization axis perpendicular to the reaction plane. θ is then the "out-of-plane" angle, measured with respect to an axis normal to the reaction plane. In the center-of-mass system, this distribution is axially symmetric around the quantization axis. When transformed to the laboratory system, this symmetry is broken, and the distribution, which is presented as a function of θ, is integrated over ϕ.

6.7 Special Programming Features

A feature that is especially important in understanding the program structure is the parallel processing of cascades with the same A, Z, E, and J. Since a major fraction of the program time is spent calculating the sum-integral for $\Gamma(E, J)$, all of the cascades at this excitation energy and spin, and with identical values of A and Z, are processed together. Thus, increasing the Monte Carlo statistics by increasing the number of cascades N does not increase the running time proportionally. Rather, the running time is found to increase approximately as $\sqrt{N}$.

Another feature to note is the limits initially imposed on the various spectra. These are defined by the EMAX and EMIN variables at the beginning of the main program. They prevent calculation of tails of spectra that are unlikely to be obtained in the Monte Carlo procedure with a small number of events. The user can, however, change these variables at his or her discretion to obtain tails of spectra. At the same time, the user should run the code with a much larger number of events to utilize these changes. The code is currently dimensioned for 9996 events. In order to increase the number of events beyond this limit, one can make a global change (with a text editor) of 9996 to the desired number of events. (This program was written before the PARAMETER statement was implemented in FORTRAN). Similarly, higher excitation energies can be run by making a global change of 286 to a larger value, to accommodate the higher energy. The level-density dimension, currently 19696, can also be changed by changing all 19696s in the program.

6.8 Input and Output Files

The main input and output are read and written to the system-defined input and output streams. In addition to these, the program needs a mass table, read in on logical unit 3. The Wapstra-1983 mass table is supplied with the code. Two output files are produced. Logical unit 2 contains binary information for each and every decay step in the cascades. This information is the source for the tables produced by subroutine `STATIS`, called at the end of the calculation. Logical unit 4 provides formatted output on level densities and optical model parameters, should the user desire more detailed information than provided in the standard output stream. Example input (`PACE2.INP`) and output (`PACE2.OUT`) files for the reaction ^{12}C on ^{197}Au at a beam energy of 100 MeV are included on the disk.

6.0 Summary

The `PACE2` code has been used by many research groups and for many different reactions over the past several years. Some examples of the use of statistical model calculations in heavy-ion reactions are presented in reference [6.13]. The code contains defaults for everything it needs to run. However, the user should be aware that the results obtained are only as good as the data used in the code. Specific applications may demand that the user provide discrete levels, select a specific optical-model potential, formulate his or her own level-density tables, improve the gamma-decay parametrization, or do anything else that is necessary to enhance the accuracy and reliability of a particular calculation.

6.10 Technical Note

Before running the code, the user should read the detailed informations given at the beginning of the source file. Here, he or she will find additional details about the program and its options as well as a detailed line-by-line explanation of the input required by the code (for comparison, see the example input file `PACE2.INP` included on the disk). As the code is not intended to be run on a microcomputer, conversion to strict ANSI-standard FORTRAN 77 has not been attempted.

References

[6.1] V. Weisskopf, Phys. Rev. **52**, 295 (1937)

[6.2] J. M. Alexander, M. T. Magda, and S. Landowne, Phys. Rev. C**42**, 1092 (1990)

[6.3] W. Hauser and H. Feshbach, Phys Rev **87**, 366 (1952)

[6.4] J. R. Grover, Phys. Rev. **127**, 2142 (1962)

[6.5] M. Hillman and Y. Eyal, unpublished

[6.6] N. Bohr and J. A. Wheeler, Phys. Rev. **56**, 426 (1939)

[6.7] A. Gilbert and A. G. W. Cameron, Can. J. Phys. **43**, 1446 (1965)

[6.8] A. J. Sierk, Phys. Rev. C**33**, 2039 (1986)

[6.9] S. Cohen, F. Plasil, and W. J. Swiatecki, Ann. Phys. (N.Y.) **82**, 557 (1974)

[6.10] C. M. Perey and F. G. Perey, At. Nucl. Data Tables **17**, 1 (1976)

[6.11] J. R. Huizenga and G. Igo, Nucl. Phys. **29**, 462 (1962)

[6.12] D. L. Hill and J. A. Wheeler, Phys. Rev. **89**, 1102 (1953)

[6.13] A. Gavron, Phys. Rev. C**21**, 230 (1980)

7. The Time-Dependent Hartree–Fock Approximation for Nuclear Slabs

J.A. Maruhn and S.E. Koonin

7.1 Background

The time-dependent Hartree–Fock (TDHF) approximation is a rare example of a nontrivial quantal many-body system whose time-dependence – within the limitations of the numerical methods employed – can be studied exactly. For this reason, it has given important insight into the reaction mechanism of heavy-ion collisions. The basic assumption behind TDHF is that the nucleons can at all times be regarded as independent particles that interact only through a mean field. Mathematically, this means that the many-body wave function always corresponds to a Slater determinant built out of a set of time-dependent single-particle orbitals; correlations or direct two-body collisions are neglected. The TDHF approximation is expected to be valid for bombarding energies per nucleon that are significantly below the Fermi energy in nuclei, $E_{\rm cm} \ll \epsilon_{\rm F} \approx 40$ MeV, where the Pauli principle prohibits most of the direct nucleon–nucleon collisions.

The most important references concerning the original calculations are given in [7.1–16], and more can be found in the review articles [7.17–19]. A good overview of Hartree–Fock methods in general and their connection to other approaches can also be found in [7.20]. The code we present here is an implementation following closely the ideas of the original work of Bonche, Koonin, and Negele [7.1].

Given the Hamiltonian for a system of A particles interacting through two-body potentials,

$$H = -\sum_i \frac{\hbar^2}{2m}\nabla_i^2 + \sum_{i\neq j} v(\boldsymbol{r}_i, \boldsymbol{r}_j), \tag{7.1}$$

with the coordinates $\boldsymbol{r}_i$ implicitly denoting also spin and isospin indices, we approximate the many-body wave function by a Slater determinant ($\mathcal{A}$ is the antisymmetrization operator):

$$\Psi(\boldsymbol{r}_1, \ldots, \boldsymbol{r}_A, t) = \mathcal{A} \prod_{n=1}^{A} \psi_n(\boldsymbol{r}_n, t). \tag{7.2}$$

Minimizing the action

$$S = \int_{t_1}^{t_2} \mathrm{d}t \langle \Psi(t)| \mathrm{i}\hbar \frac{\partial}{\partial t} - H |\Psi(t)\rangle \tag{7.3}$$

leads to the TDHF equations

$$\begin{aligned} i\hbar\frac{\partial}{\partial t}\psi_n(\boldsymbol{r},t) &= -\frac{\hbar^2}{2m}\nabla^2\psi_n(\boldsymbol{r},t) \\ &+\sum_{j=1}^{A}\int \mathrm{d}^3r'\, v(\boldsymbol{r}-\boldsymbol{r}')|\psi_j(\boldsymbol{r}',t)|^2\,\psi_n(\boldsymbol{r},t) \\ &-\sum_{j=1}^{A}\int \mathrm{d}^3r'\, v(\boldsymbol{r}-\boldsymbol{r}')\psi_j^*(\boldsymbol{r}',t)\psi_n(\boldsymbol{r}',t)\,\psi_j(\boldsymbol{r},t). \end{aligned} \tag{7.4}$$

These correspond to the static Hartree–Fock equations, with the time derivative appearing in the same way as in the Schrödinger equation.

The time development of the system thus is governed by a set of A nonlinear integro-differential equations coupled by the average potential. The numerical complexity is considerably reduced if the exchange term can be converted to a local form, which is true for zero-range interactions like the *Skyrme forces* [7.21], so that almost all calculations to date have employed this type of force (an exception is given in [7.2], where a G-matrix is used). The density dependence of these interactions poses no problems.

If there is only a local term, the TDHF equations reduce to

$$i\hbar\frac{\partial}{\partial t}\psi_n(\boldsymbol{r},t) = -\frac{\hbar^2}{2m}\nabla^2\psi_n(\boldsymbol{r},t) + W(\boldsymbol{r})\psi_n(\boldsymbol{r},t), \tag{7.5}$$

with $W(\boldsymbol{r},t)$ the *mean field* resulting from the evaluation of the corresponding terms in (7.4). The mean field depends on the wave functions via only averaged quantities, such as the total density or total current density.

The *initial conditions* describe two separated "nuclei" in their respective ground states with a given relative motion. The ground-state wave functions for each nucleus are obtained from the solution of the corresponding static Hartree–Fock problem, that is, (7.5) with $i\hbar\partial/\partial t$ replaced by the single-particle energy ϵ_n, and with the average potential computed from the density for that nucleus only. These static wave functions are multiplied by the appropriate plane-wave factor to start the relative motion.

7.2 Interaction

For the reasons given above, the code presented here employs an interaction of Skyrme type. However, the general Skyrme force

$$\begin{aligned} v(\boldsymbol{r},\boldsymbol{r}') &= t_0\delta(\boldsymbol{r}-\boldsymbol{r}') + \tfrac{1}{2}t_1[\delta(\boldsymbol{r}-\boldsymbol{r}')k'^2 + k^2\delta(\boldsymbol{r}-\boldsymbol{r}')] \\ &+t_2\boldsymbol{k}\cdot\delta(\boldsymbol{r}-\boldsymbol{r}')\boldsymbol{k}' + \tfrac{1}{6}t_3\rho(\boldsymbol{r})\delta(\boldsymbol{r}-\boldsymbol{r}'), \end{aligned} \tag{7.6}$$

with $\rho(\boldsymbol{r})$ the local density of nucleons, has the disadvantage that the derivatives in the t_1 and t_2 terms will couple the different spatial directions and would make it impossible to treat a slab in a purely one-dimensional manner. One can instead use a Yukawa interaction, so that the force actually used in the code is

$$v(\boldsymbol{r},\boldsymbol{r}') = t_0\delta(\boldsymbol{r}-\boldsymbol{r}') + \tfrac{1}{6}t_3\rho\delta(\boldsymbol{r}-\boldsymbol{r}') + W_0\frac{\mathrm{e}^{-|\boldsymbol{r}-\boldsymbol{r}'|/a}}{|\boldsymbol{r}-\boldsymbol{r}'|/a}P. \tag{7.7}$$

The Yukawa potential of course leads to an unwanted non-local exchange term; for this reason, there is the additional exchange operator P whose sole purpose it is to make the Yukawa part contribute only to the direct term. Its actual form is given in [7.1], but is really of no importance.

In the code, the assumption of a *spin and isospin saturated system* is also made, so that each spatial wave function is occupied by four nucleons, which are coupled to vanishing total spin and isospin. Evaluating the direct and exchange terms of the mean field with this assumption is quite straightforward and illustrated here for the t_0-term. Remembering that the integration over $\boldsymbol{r}$ also implies summation over spin and isospin degrees of freedom, and expressing these degrees of freedom explicitly in the wave functions

$$\psi_n(\boldsymbol{r},t) \to \psi_n(\boldsymbol{r},t)\chi_n(\sigma)\eta_n(\tau), \tag{7.8}$$

with χ and η the spin and isospin wave functions, the direct term can be written as

$$\begin{aligned} t_0 \ \textstyle\sum_j & \int \mathrm{d}^3r' \sum_{\sigma'\tau'} \delta(\boldsymbol{r}-\boldsymbol{r}')|\psi_j(\boldsymbol{r}')\chi_j(\sigma')\eta_j(\tau')|^2\psi_n(\boldsymbol{r})\chi_n(\sigma)\eta_n(\tau) \\ &= t_0\rho(\boldsymbol{r})\psi_n(\boldsymbol{r})\chi(\sigma)\eta(\tau), \end{aligned} \tag{7.9}$$

while the exchange term reduces to

$$\begin{aligned} -t_0 \ \textstyle\sum_j & \int \mathrm{d}^3r' \sum_{\sigma'\tau'} \delta(\boldsymbol{r}-\boldsymbol{r}')\psi_j^*(\boldsymbol{r}')\chi_j(\sigma')\eta_j(\tau')\psi_n(\boldsymbol{r}')\chi_n(\sigma')\chi_n(\tau') \\ & \times\psi_j(\boldsymbol{r})\chi_j(\sigma)\eta_j(\tau) \\ &= -t_0\sum_j |\psi_j(\boldsymbol{r})|^2\psi_n(\boldsymbol{r})\chi_j(\sigma)\eta_j(\tau)\delta_{\sigma_j\sigma_n}\delta_{\tau_j\tau_n} \\ &= -\tfrac{1}{4}t_0\rho(\boldsymbol{r})\psi_n(\boldsymbol{r})\chi_n(\sigma)\eta_n(\tau), \end{aligned} \tag{7.10}$$

because the Kronecker symbols in σ and τ pick out only those of the wave functions ψ_j whose spin and isospin agree with that of the wave function ψ_n. Working out similar expressions for the other terms and neglecting the exchange term for the Yukawa potential because of the operator P, we end up with the mean-field

$$\begin{aligned} W(\boldsymbol{r}) &= \tfrac{3}{4}t_0\rho(\boldsymbol{r}) + \tfrac{3}{16}t_3\rho^2(\boldsymbol{r}) \\ & + W_0\int \rho(\boldsymbol{r}')\frac{\mathrm{e}^{-|\boldsymbol{r}-\boldsymbol{r}'|/a}}{|\boldsymbol{r}-\boldsymbol{r}'|/a}\mathrm{d}\boldsymbol{r}'. \end{aligned} \tag{7.11}$$

Suggested values for the potential parameters are given in [7.1]:

$t_0 = -497.726$ MeV fm^3,
$t_3 = 17270$ MeV fm^6,
$a = 0.45979$ fm,
$W_0 = 363.0438$ MeV.

7.3 Reduction to One Dimension

Rather than taking the simplest approach of true one-dimensional motion, one may hope to retain some more physics through the use of *slab geometry*; that is, infinite homogeneous nuclear matter in both the y and z directions, with the wave functions, density, and mean potential depending only on the x-coordinate. We consider the static problem first.

The time-independent single-particle equation

$$-\frac{\hbar^2}{2m}\nabla^2\psi_i(\boldsymbol{r}) + W(x)\psi_i(\boldsymbol{r}) = \epsilon_i\psi_i(\boldsymbol{r}) \tag{7.12}$$

is separable, with the transverse parts given by plane waves:

$$\psi_i(\boldsymbol{r}) = \psi_{n,k_yk_z}(\boldsymbol{r}) = \frac{1}{\sqrt{\Omega}}\exp[\mathrm{i}(k_y y + k_z z)]\ \phi_n(x). \tag{7.13}$$

Here Ω denotes the normalization area in (y, z), and $\phi_n(x)$, the dynamically varying wave functions. The index i was replaced by the triplet n, k_y, k_z. The ϕs are normalized to

$$\int_{-\infty}^{\infty} |\phi_n(x)|^2\, \mathrm{d}x = 1. \tag{7.14}$$

The eigenvalue is similarly decomposed as

$$\epsilon_i = \epsilon_n + \frac{\hbar^2}{2m}(k_y^2 + k_z^2), \tag{7.15}$$

where ϵ_n results from the solution of the one-dimensional Schrödinger-like equation

$$-\frac{\hbar^2}{2m}\frac{\partial^2\phi_n(x)}{\partial x^2} + W(x)\phi_n(x) = \epsilon_n\phi_n(x), \tag{7.16}$$

with the potential given by

$$W(x) = \tfrac{3}{4}t_0\rho(x) + \tfrac{3}{16}t_3\rho^2(x) + 2\pi a^2 W_0 \int \mathrm{d}x'\rho(x')\mathrm{e}^{-|(x-x')|/a}. \tag{7.17}$$

The presence of the transverse degrees of freedom is important only when occupying the eigenstates up to the Fermi energy ϵ_F according to equation (7.15); this implies that for a given energy in the x-direction, ϵ_n, all transverse states with

$$k_\mathrm{max}(n)^2 = k_y^2 + k_z^2 \le \frac{2m}{\hbar^2}(\epsilon_\mathrm{F} - \epsilon_n) \tag{7.18}$$

are occupied. This may be taken into account by introducing fractional occupation numbers for the states $\phi_n(x)$ as follows. The total density is given by the sum over all occupied states, with a factor of 4 taking account of the fourfold occupation of the states due to spin and isospin degeneracy:

$$\begin{aligned}\rho(x) &= 4\sum_{n,k_y,k_z} |\psi_{n,k_y,k_z}(\boldsymbol{r})|^2 \\ &= 4\sum_n |\phi_n(x)|^2 \int_0^{k_{\max}(n)} \frac{\mathrm{d}^2k}{(2\pi)^2} \\ &= \sum_n |\phi_n(x)|^2 \frac{k_{\max}^2(n)}{\pi}. \end{aligned} \tag{7.19}$$

If occupation numbers A_n are defined as

$$A_n = \frac{k_{\max}^2(n)}{\pi} = \frac{2m}{\hbar^2}\frac{\epsilon_{\mathrm{F}} - \epsilon_n}{\pi}, \tag{7.20}$$

the total density can be written as

$$\rho(x) = \sum_n A_n |\phi_n(x)|^2, \tag{7.21}$$

and the total number of particles becomes, because of the normalization of the $\phi_n(x)$,

$$A = \sum_n A_n = \sum_{n=1}^{N} \frac{2m}{\hbar^2\pi}(\epsilon_{\mathrm{F}} - \epsilon_n), \tag{7.22}$$

with N the total number of occupied one-dimensional states. Note that now the problem has been reduced to a one-dimensional form while retaining the three-dimensional interaction and phase space; this is possible only for such simple interactions that allow a separation of the single-particle equations.

For the time-dependent problem, the situation is similar. Because the mean potential only depends on the x-coordinate, the transverse components of the wave function remain time independent – aside from the trivial phase factor $\exp[-\mathrm{i}\hbar(k_y^2 + k_z^2)t/2m]$. Thus, we end up with the time-dependent equation

$$\begin{aligned}\mathrm{i}\hbar\frac{\partial}{\partial t}\phi_n(x,t) &= \left[-\frac{\hbar^2}{2m}\frac{\partial^2}{\partial x^2} + W(x,t)\right]\phi_n(x,t) \\ &\equiv H(t)\phi_n(x,t). \end{aligned} \tag{7.23}$$

Here, the single-particle Hamiltonian H depends on time implicitly through the dependence of the mean-field W on the wave functions [that is, in our case simply on the density, see (7.11)]. The occupation numbers A_n are time-independent because of the complete decoupling of the transverse motion.

7.4 Numerical Methods

7.4.1 The Static Problem

To determine the eigenfunctions in a static nucleus, we use a very simple iteration, which is satisfactory mainly because computation times are not excessive and the static problem has to be solved only infrequently. Assume that the true eigenfunctions of the single-particle Hamiltonian H are $\phi_n(x)$ with eigenvalues ϵ_n, and that $\psi(x)$ is a trial wave function whose expansion in the eigenstates is

$$\psi(x) = \sum_n a_n \phi_n(x). \tag{7.24}$$

An iteration step of the form

$$\psi' \equiv [1 - \Delta t(H - E_0)/\hbar]\psi = \sum_n a_n[1 - (\epsilon_n - E_0)\Delta t/\hbar]\phi_n(x) \tag{7.25}$$

will reduce the admixture of higher-energy wave eigenstates, provided that the time step Δt is chosen such that $\Delta t(\epsilon_n - E_0) < 2\hbar$ for all n.

On a spatial grid x_j, $j = 0, \ldots, N$, the action of H is approximated with second-order differencing. Denoting $\psi(x_j) = \psi_j$, it can be written as

$$\psi_j' = [1 - \Delta t(W_j - E_0)/\hbar]\psi_j + \frac{\hbar^2}{2m\Delta x^2}(\psi_{j+1} - 2\psi_j + \psi_{j-1}). \tag{7.26}$$

Since the iteration does not conserve normalization, the wave function has to be normalized after each iteration.

To solve for excited states in addition to the ground state, one only has to iterate several wave functions simultaneously and explicitly orthogonalize them after each iteration. As initial wave functions, the code uses the eigenstates of the one-dimensional harmonic oscillator.

The occupation numbers have to be adjusted because of changes in the single-particle energies during the iteration. This can be done simply using (7.22). Assume that N states are occupied, then calculate ϵ_F from (7.22). The result is satisfactory if the Fermi energy fulfills

$$\epsilon_N \le \epsilon_\mathrm{F} \le \epsilon_{N+1}.$$

If this is not true, increase N by one and try again.

The selection of the proper time step Δt is a crucial problem. The maximum energy above the potential minimum expected in the system can be estimated as the maximal kinetic energy of a plane wave on the given grid, which is given by

$$\begin{aligned} T(k) &= -\frac{\hbar^2}{2m\Delta x^2}(\mathrm{e}^{ik\Delta x} - 2 + \mathrm{e}^{-ik\Delta x}) \\ &= \frac{\hbar^2}{2m\Delta x^2}(2 - 2\cos k\Delta x) \\ &\le \frac{2\hbar^2}{m\Delta x^2}. \end{aligned} \tag{7.27}$$

The maximal time step is, thus, given by

$$\Delta t_{\max} = \frac{m\Delta x^2}{\hbar} \tag{7.28}$$

or $\Delta t_{\max}(\mathrm{fm}/c) \approx 4.7\,[\Delta x(\mathrm{fm})]^2$.

The parameter E_0 for fastest convergence should be close to the expected lowest single-particle energy. In the present code, it was for simplicity chosen to be the minimum of the average potential.

The code also employs a simple relaxation: in each iteration, the mean field and the wave functions are averaged with those from the preceding time step to damp oscillatory convergence. This is controlled by a variable "DAMP": the old values are weighted with a factor of DAMP and the new ones with 1 − DAMP.

A special problem is the evaluation of the Yukawa integral necessary for the mean field. A straightforward application of Simpson's rule yields too inaccurate results, because the range of the Yukawa interaction is comparable with the grid spacing. An improved method based on a higher-order approximation is given in [7.1] (Appendix A) and faithfully implemented in the present code.

7.4.2 The Dynamic Problem

Once the static wave functions have been obtained, they are used to initialize the time-dependent problem. The wave functions are inserted into the grid a distance "DIST" apart and multiplied with a plane-wave phase factor $\exp(\pm ikx)$, where $E_{\mathrm{coll}} = \hbar^2k^2/2m$, and the sign is chosen for the projectile and target nucleons in such a way that they start moving towards each other.

For the numerical solution of the time-dependent problem, the basic idea is to approximate the evolution operator for a short time step Δt. Formally, the solutions to the time-dependent Hartree–Fock equations are given by

$$\phi(x, t + \Delta t) = \exp\left[\frac{\mathrm{i}}{\hbar}\int_t^{t+\Delta t} \mathrm{d}t'\; H(t')\right]\phi(x, t). \tag{7.29}$$

The time dependence of the single-particle Hamiltonian poses a special problem, since it depends on the unknown wave functions. It is not sufficient to just approximate it by its value at the beginning of the time step; in practice, it was found that this procedure violates energy conservation strongly. For example, for the case of a nucleus moving uniformly through space, the mean field in the Hamiltonian would always correspond to the initial position of the nucleus during the time step and thus lag behind, causing a spurious slowing of the motion. Satisfactory accuracy can be reached if the Hamiltonian is estimated at the center of the time step (the same numerical effect is also illustrated by the midpoint approximation to an integral). The estimation itself can be done with lower accuracy.

The propagation for one time step thus proceeds in two phases:

- Estimation of $H(t+\Delta t/2)$: estimate the wave functions at half-time via
$$\phi(x,t+\Delta t/2) = \exp\left[\frac{\mathrm{i}}{\hbar}H(t)\Delta t/2\right]\phi(x,t), \tag{7.30}$$
and then compute the density and the Hamiltonian according to (7.23).
- In the second phase, use this Hamiltonian for the final propagation
$$\phi(x,t+\Delta t) = \exp\left[\frac{\mathrm{i}}{\hbar}H(t+\Delta t/2)\Delta t\right]\phi(x,t). \tag{7.31}$$
These wave functions then provide the starting point for the next time step.

The exponential of the Hamiltonian in (7.30) and (7.31) is approximated in a manifestly unitary way through

$$\exp(\mathrm{i}H\Delta t/\hbar) \approx \frac{1-\mathrm{i}H\Delta t/2\hbar}{1+\mathrm{i}H\Delta t/2\hbar}. \tag{7.32}$$

A unitary approximation is highly desirable because it automatically maintains the normalization of the wave functions, preventing one source of instability. Non-unitary methods can also be used, but have to be highly accurate to avoid instabilities leading to a sudden "explosion" in the norms.

It is slightly more efficient to rewrite (7.32) in its action on the wave functions as

$$\phi(x,t+\Delta t) = \left(\frac{2}{1+\mathrm{i}H\Delta t/2\hbar} - 1\right)\phi(x,t) \equiv \chi(x) - \phi(x,t). \tag{7.33}$$

The intermediate function $\chi(x)$ has to be determined from

$$(1+\mathrm{i}H\Delta t/2\hbar)\,\chi(x) = 2\phi(x,t). \tag{7.34}$$

We now have to write this equation in discretized form. Introducing the abbreviations $\phi(j\Delta x, n\Delta t) \equiv \phi_j^n$, and $\chi(j\Delta x) \equiv \chi_j$, and using the standard three-point formula for the second derivative in the kinetic energy, we get

$$\chi_j + \frac{\mathrm{i}\Delta t}{2\hbar}\left[-\frac{\hbar^2}{2m}\frac{1}{\Delta x^2}(\chi_{j+1} - 2\chi_j + \chi_{j-1}) + W_j\right] = 2\phi_j^n. \tag{7.35}$$

Collecting terms and dividing by a factor such that the coefficients of $\chi_{j\pm1}$ become unity, the resulting equation is

$$\chi_{j+1} + \left(-2 + \frac{4\mathrm{i}m\Delta x^2}{\hbar\Delta t} - \frac{2m\Delta x^2}{\hbar^2}W_j\right)\chi_j + \chi_{j-1} = \frac{8\mathrm{i}m\Delta x^2}{\hbar\Delta t}\phi_j^n. \tag{7.36}$$

This system of linear equations has the general form

$$A_j^-\chi_{j-1} + A_j^0\chi_j + A_j^+\chi_{j+1} = b_j, \tag{7.37}$$

which may be solved by the *two-sweep method.* We assume that the solution is given by a one-term forward recursion relation of the form

$$\chi_{j+1} = \alpha_j \chi_j + \beta_j, \tag{7.38}$$

where the coefficients α_j and β_j have to be determined. Substituting the recursion relation into (7.37), we get

$$A_j^- \chi_{j-1} + A_j^0 \chi_j + A_j^+ (\alpha_j \chi_j + \beta_j) = b_j, \tag{7.39}$$

which can be solved for χ_j to yield

$$\chi_j = \gamma_j A_j^- \chi_{j-1} + \gamma_j (A_j^+ \beta_j - b_j), \tag{7.40}$$

with

$$\gamma_j = -\frac{1}{A_j^0 + A_j^+ \alpha_j}. \tag{7.41}$$

Comparing the coefficients with the recursion relation (7.38), we obtain

$$\alpha_{j-1} = \gamma_j A_j^-, \quad \beta_{j-1} = \gamma_j (A_j^+ \beta_j - b_j). \tag{7.42}$$

Thus, to solve the linear system (7.37), one has to compute the α_j and β_j from these backward recursion relations and afterwards the χ_j from the forward recursion. All that remains is to fix the starting values for the recursion. If we take boundary conditions such that all the wave functions vanish at $j = 0$ and $j = N$, we should have $\alpha_{N-1} = \beta_{N-1} = 0$. This algorithm is implemented in the code.

The same method is used for computing both the estimate for the half-time wave functions and the final propagated values. A fuller account of the numerical method may be found in Chapter 7 of [7.22].

7.5 Remarks on the Code

The code is a straightforward implementation of the algorithms outlined above. Most of the values are passed in COMMON, with only those in argument lists that actually are different in different calls of the same routine. The dimension statements all use parameter variables to indicate the maximal dimensions allowed, while for the actual calculation any dimension smaller than that may be selected.

To run a collision, the user first has to obtain the static wave functions for the two participating nuclei in a static calculation. They are written into a file and may be given as part of the input for a dynamic calculation.

7.5.1 Input

Control of the program requires only three input lines. The first and third are the same for the static and dynamic cases, with the second taking care of the different requirements. The only other values that may need changing are the parameters NMAX for the maximum number of wave functions and

Table 7.1. Input lines for the TDHF code. On each line the variables listed have to be entered in free-field format in the exact order given. The variable types are also indicated.

Line 1: *(static and dynamic)*		
STATIC	logical	.TRUE. for static calculation, .FALSE. for dynamic case.
NP	integer	number of points in space grid.
ITMAX	integer	maximum number of iterations or time steps.
IPLOT	integer	frequency of statistics and printer plots.
DT	double	time step in fm/c.
DX	double	spatial grid spacing in fm.
Line 2: *(dynamic case only)*		
FNAME1	character*30	file name containing static result for projectile.
FNAME2	character*30	file name containing static result for target.
ECOLL	double	kinetic energy per nucleon in MeV for both projectile and target.
DIST	double	initial distance between the nuclei in fm.
Line 2': *(static case only)*		
FNAME1	character*30	name of the file in which the results of the calculation will be written for later input to a dynamic calculation.
ATOT	double	total "mass number" of the desired nucleus in fm^{-2}.
ERR	double	convergence criterion. The calculation is stopped when the sum of the absolute changes in the single-particle energies, multiplied by their occupation number, is less than ERR.
N	integer	number of states to be iterated.
DAMP	double	weighting of the previous mean field and wave functions for damped iteration.
Line 3: *(both static and dynamic)*		
T0	double	Skyrme-force parameter t_0 in MeV fm^3.
T3	double	Skyrme-force parameter t_3 in MeV fm^6.
A	double	Yukawa range parameter a in fm.
V0	double	Yukawa strength W_0 in MeV.

NPMAX for the maximum number of spatial grid points (more precisely, the associated index runs from 0 through NPMAX).

For a complete definition of the input refer to Table 7.1. **Important:** The static calculation has to be consistent with the dynamic one with respect to the force parameters T0, T3, A, V0, and the grid spacing DX. The code does not check this consistency and, in particular, does not perform interpolation to a different grid spacing!

7.5.2 Output

The code regularly – at intervals determined by the value of IPLOT – prints out summary statistics about the single-particle wave functions and the total system. We first discuss the single-particle quantities by indicating what precisely is printed under the appropriate header.

- Norm: the norm of the wave function, that is, $\int |\phi_n|^2 \, \mathrm{d}x$. This is useful primarily for checking the stability and accuracy of the calculation.
- Energy: the expectation value of the single-particle Hamiltonian.
- Center: the expectation value of x for this wave function, defined as
  $$\bar{x}_n = \int_{-\infty}^{\infty} \mathrm{d}x \, x |\phi_n(x)|^2.$$
  It shows the average position, but may lose information value after the collision when wave functions tend to fragment.
- Sigma: mean-square deviation of x. Gives an idea of the dispersal of the wave function during the collision,
  $$\sigma_n = \int_{-\infty}^{\infty} \mathrm{d}x \, (x - \bar{x}_n)^2 |\phi_n(x)|^2.$$
- Aleft: fraction of the probability distribution to the left of the origin,
  $$F_n^< = \int_{\infty}^{x_c} \mathrm{d}x \, |\phi_n(x)|^2.$$
  In a collision, it should be zero or one initially, and after the collision, it indicates to what extent the wave function is being shared by fragments moving left or right.

Similar quantities for the total system are defined by summing up with proper accounting for the occupation numbers; thus,

$$\bar{x} = \sum_n A_n x_n, \qquad \sigma = \sum_n A_n \sigma_n, \qquad F^< = \frac{1}{A} \sum_n A_n F_n^<. \tag{7.43}$$

The total mass per unit area is also printed out, while the total energy per unit area has to include the usual correction of the potential contribution (see any textbook on Hartree–Fock methods, for example [7.20]), and the contribution of the plane waves in the transverse degrees of freedom:

$$\begin{aligned} E &= \frac{\hbar^2}{2m} \sum_{n=1}^{N} A_n \int \left| \frac{\mathrm{d}\phi_n(x)}{\mathrm{d}x} \right|^2 \mathrm{d}x + \sum_{n=1}^{N} \frac{\hbar^2 \pi A_n^2}{4m} \\ & + \frac{3}{8} \int t_0 \rho^2(x) \mathrm{d}x + \frac{t_3}{16} \int \rho^3(x) \mathrm{d}x \end{aligned}$$

$$+\pi a^2 W_0 \int \mathrm{d}x \int \mathrm{d}x' \rho(x)\rho(x') \exp(-|x-x'|/a). \tag{7.44}$$

The program also prints out a simple printer plot. This routine "PLOT" was included because of the lack of a standard for graphics in FORTRAN and may be replaced by locally available fancier graphics packages if desired.

7.6 Running the Code

The code is contained in the file "`TDHF.FOR`" and does not require any non-standard subroutine libraries. It should simply be compiled, linked, and run. Basic input and output use the default unit "*"; they may be redirected or the few READ and WRITE statements in the code may be changed.

A test case is provided giving sample input and output for both the static and dynamic cases. The input files are "`STATIC.INP`" and "`DYNAM.INP`," and the corresponding output files differ by having an extension of "`OUT`." The static input produces wave functions for a system with a width corresponding to the size of a real ^{208}Pb nucleus, written onto a file named "`PB208.DAT`," while the dynamic case runs a collision of two such systems at 10 MeV per nucleon.

Note that the static calculation needs a large number of iterations (more than 1000) because of the relative simplicity of both the algorithm and the initialization, and because of the occasional change in the set of occupied single-particle wave functions. In this case, the most rapid convergence was found for DAMP set to zero.

7.7 Things to Do

It is a good idea to read the original paper [7.1] to get an impression of the types of behavior one might expect as well as reasonable ranges for the parameters. Aside from this, a list of interesting and useful projects should include:

- Numerical stability: how to choose the spatial and temporal discretization. What happens if Δx or Δt is too large in the static and dynamic cases?
- An investigation of the static solutions in their dependence on nucleon number and interaction.
- In the dynamic case, there are many interesting properties that may be studied systematically. How does the outcome of the reaction depend on energy or target and projectile sizes? Do reaction channels such as fusion and multifragmentation depend on energy in a simple "threshold" way or do they occur intermixed throughout some interval? How does the exact choice of the interaction affect the collision mechanism? Is there transparency at high energies?

References

[7.1] P. Bonche, S. E. Koonin, and J. W. Negele, Phys. Rev. **13**, 1226 (1976)
[7.2] J. A. Maruhn and R. Y. Cusson, Nucl. Phys. A**270**, 471 (1976)
[7.3] R. Y. Cusson, R. K. Smith, and J. A. Maruhn, Phys. Rev. Lett. **36**, 1166 (1976)
[7.4] C. Y. Wong, J. A. Maruhn, and T. A. Welton, Phys. Lett. **66B**, 19 (1977)
[7.5] R. Y. Cusson, J. A. Maruhn, and H. W. Meldner, Phys. Rev. C**18**, 2589 (1978)
[7.6] K. R. Sandhya-Devi and M. R. Strayer, Phys. Lett. **77B**, 135 (1978)
[7.7] K. T. R. Davies, V. Maruhn-Rezvani, S. E. Koonin, and J. W. Negele, Phys. Rev. Lett. **41**, 632 (1978)
[7.8] K. T. R. Davies, H. T. Feldmeier, H. Flocard, and M. S. Weiss, Phys. Rev. C**18**, 2631 (1978)
[7.9] K. T. R. Davies, K. R. Sandhya Devi, and M. R. Strayer, Phys. Rev. C**20**, 1372 (1979)
[7.10] P. Bonche et al., Phys. Rev. C**20**, 641 (1979)
[7.11] K. T. R. Davies, K. R. Sandhya Devi, and M. R. Strayer, Phys. Rev. Lett. **44**, 23 (1980)
[7.12] R. Y. Cusson, J. A. Maruhn, and H. Stöcker, Z. Physik A**294**, 257 (1980)
[7.13] K. T. R. Davies, K. R. Sandhya Devi, and M. R. Strayer, Phys. Rev. C**24**, 2576 (1981)
[7.14] K. T. R. Davies and S. E. Koonin, Phys. Rev. C**23**, 2042 (1981)
[7.15] J. A. Maruhn, K. T. R. Davies, and M. R. Strayer, Phys. Rev. C**31**, 1289 (1985)
[7.16] A. S. Umar et al., Phys. Rev. C**32**, 172 (1985)
[7.17] J. W. Negele, Rev. Mod. Phys. **54**, 913 (1982)
[7.18] K. T. R. Davies, K. R. Sandhya Devi, S. E. Koonin, and M. R. Strayer, in *Heavy Ion Science, Nuclear Science*, edited by D. A. Bromley (Plenum, New York, 1985) Vol. 3, p. 3.
[7.19] P. Mädler, *Fizika Elementarnykh Chastits i Atomnogo Yadra* **15**, 418 (1984)
[7.20] P. Ring and P. Schuck, *The Nuclear Many-Body Problem* (Springer-Verlag, New York, 1980)
[7.21] T. H. R. Skyrme, Phil. Mag. **1**, 1043 (1956), Nucl. Phys. **9**, 615 (1959)
[7.22] S. E. Koonin, *Computational Physics* (Benjamin/Cummings, Menlo Park, 1986)

8. The Vlasov–Uehling–Uhlenbeck Model

C. Hartnack, H. Kruse, and H. Stöcker

8.1 Introduction

Little is known about the properties of nuclear matter under extreme conditions, that is, at high temperatures and high compression, a state that might have existed in the early universe. The only knowledge we have is the behavior of normal nuclear matter at ground-state density $\varrho = \varrho_0 \approx 0.15 - 0.17$ fm^{-3} and temperature $T = 0$.

In order to extract information about hot and dense nuclear matter, there are two possibilities. One is to study supernovae and the inner part of neutron stars, where high densities and rather moderate temperatures are achieved. However, a lot of knowledge about the structure and chemical composition of stars and about the mechanism of collapse is necessary. Furthermore, the number of visible supernovae and of radiating neutron stars is finite.

The other possibility is the study of high-energy heavy-ion collisions [8.1]. The key mechanism is the formation of shock waves suggested by the nuclear fluid-dynamical model [8.2]. These occur if the velocity of the incoming projectile is larger than the velocity of sound in equilibrium nuclear matter [8.2],

$$v > c_s \approx 0.2c. \tag{8.1}$$

As a consequence of this mechanism, heavy colliding systems are stopped to their center-of-mass rapidity and hot high-energy particles are ejected perpendicularly to the beam axis for central, that is, head-on-head, heavy-ion collisions. For semicentral collisions, particles are squeezed out perpendicularly to the reaction plane (determined by beam axis and impact parameter) and projectile and target remnants are scattered off the high-density region and flow sidewards in the reaction plane. Many of these effects have been observed experimentally [8.3].

This chapter presents a model for the description of heavy-ion collisions, the so-called Vlasov–Uehling–Uhlenbeck (VUU) model. After some motivating remarks about models describing heavy-ion collisions, we will explain the basic transport equation, the so-called VUU-equation, and discuss the idea of the nuclear equation of state. Afterwards, we will discuss the numerical realization and the basic structure of the program. Finally, we will give some explanations about the use of the program.

8.2 Theoretical Description of Heavy-Ion Collisions

The description of heavy-ion collisions may be done at a macroscopic or microscopic level. In a macroscopic description, the distribution function f of the total system is propagated using transport equations and conservation laws. Only the interesting degrees of freedom are considered (for example, total energy, total momentum, and baryon number). One example of a macroscopic theory is nuclear fluid dynamics [8.2,4], a model guided by the liquid-drop model for nuclei, assuming nuclear matter to have the behavior of a compressible viscous fluid. This model succeeded in the prediction of nuclear stopping, squeeze-out, and sidewards flow (bounce off), which have been confirmed experimentally as mentioned in the introduction.

At the microscopic level, the distribution function f is obtained by summation over all distribution functions f_i of the single particles i, which in general might be quantal Wigner functions. VUU, however, assumes δ-functions in coordinate and momentum space, corresponding to the approximation by classical point particles. These particles are propagated according to classical equations of motion, and the propagated distribution function f is represented by the sum of the f_i of the propagated particles i. An advantage of this description is a more adequate treatment of binary interactions like collisions and particle production. Also, the treatment of fluctuations, for example, cluster formation, can be done at a better level. Unfortunately, the exact treatment at a full N-body level requires the solution of an N-body transport equation, which may become very difficult for large systems. Therefore, the particle interactions are reduced to a two- or three-body level in a mean-field approximation.

The two-body interactions are split up into a short-range and a long-range part as we will see in the discussion of the VUU-equation. The short-range interactions are regarded as basically corresponding to the hard-core binary collisions. Intranuclear cascade models [8.5] only deal with this short-range term. The long-range interactions are assumed to be given by a potential; commonly, a density-dependent mean field (formed by the surrounding nuclear matter) of the particle is used [8.6].

The solution of the transport equations is not done analytically but in the form of Monte Carlo simulations. One starts with the ground states of the two colliding nuclei (for example, a Fermi gas) and propagates all single nucleons, forming a statistical sample of a large number of calculations with different randomly varying ground-state configurations.

8.3 The VUU Equation

There are a number of different models based on the very same transport equation. Consequently, the equation itself is also known under different names, for example, the Vlasov–Uehling–Uhlenbeck (VUU) [8.7], Boltzmann–Uehling–Uhlenbeck [8.8], Landau–Vlasov [8.9], Vlasov–Nordheim, Landau–Vlasov–Nordheim, and Nordheim equations (probably, this list is not exhaustive). In the following, we will call it the VUU-equation.

The VUU-equation is a differential equation for the classical one-body phase-space distribution function $f(\boldsymbol{r},\boldsymbol{p},t)$ corresponding to the classical limit of the Wigner function. The equation states that the distribution function f changes in time owing to collisions only,

$$\frac{\mathrm{d}f}{\mathrm{d}t} = \left(\frac{\mathrm{d}f}{\mathrm{d}t}\right)_{\mathrm{coll}} . \tag{8.2}$$

For the left-hand side, we write down the total derivative with respect to time and assume that the forces can be derived from a potential:

$$\begin{aligned} \frac{\mathrm{d}f}{\mathrm{d}t} &= \frac{\partial f}{\partial t} + \frac{\mathrm{d}\boldsymbol{r}}{\mathrm{d}t}\frac{\partial f}{\partial \boldsymbol{r}} + \frac{\mathrm{d}\boldsymbol{p}}{\mathrm{d}t}\frac{\partial f}{\partial \boldsymbol{p}} \\ &= \frac{\partial f}{\partial t} + \boldsymbol{v}\cdot\nabla_r f - \nabla_r U\cdot\nabla_p f. \end{aligned} \tag{8.3}$$

For the collision term, only binary collisions are considered. We use the Boltzmann ansatz with the modifications given by Nordheim, Uehling, and Uhlenbeck [8.10]:

$$\begin{aligned} \left(\frac{\mathrm{d}f}{\mathrm{d}t}\right)_{\mathrm{coll}} &= -\int \frac{\mathrm{d}^3p_2\,\mathrm{d}^3p_1'\,\mathrm{d}^3p_2'}{(2\pi)^6}\sigma v_{12} \\ &\times [f f_2(1-f_1')(1-f_2') - f_1'f_2'(1-f)(1-f_2)] \\ &\times \delta^3(p+p_2-p_1'-p_2'). \end{aligned} \tag{8.4}$$

As is familiar from the theory of the Boltzmann equation, f_2' stands for $f(\boldsymbol{r},\boldsymbol{p}_2',t)$ and correspondingly for the other terms. The terms proportional to $f f_2$ and $f_1'f_2'$ represent the phase-space densities of the loss and gain terms, respectively. These terms together with the cross section σ and the relative velocity v_{12} give the transition rates for the reaction $(\boldsymbol{p},\boldsymbol{p}_2)\rightarrow(\boldsymbol{p}',\boldsymbol{p}_2')$ and for the inverse reaction. The influence of the Pauli principle is taken into account by introducing the Nordheim–Uehling–Uhlenbeck terms $(1-f_1')(1-f_2')$ and $(1-f)(1-f_2)$. These terms ensure that collisions cannot fill phase-space regions with $f\geq 1$. Finally, the δ-function ensures the conservation of energy and momentum.

The VUU-equation thus takes the form

$$\begin{aligned} \frac{\partial f}{\partial t} + \boldsymbol{v}\cdot\nabla_r f - \nabla_r U\cdot\nabla_p f &= -\int \frac{\mathrm{d}^3p_2\,\mathrm{d}^3p_1'\,\mathrm{d}^3p_2'}{(2\pi)^6}\sigma v_{12} \\ \times\ [f f_2(1-f_1')(1-f_2') - f_1'f_2'(1-f)(1-f_2)] & \\ \times\ \delta^3(p+p_2-p_1'-p_2'). & \end{aligned} \tag{8.5}$$

The two main ingredients in the VUU-equation are:

1. The potential U on the left-hand side of the VUU-equation. We use a Skyrme ansatz where different parameter sets lead to different equations of state for infinite nuclear matter having arbitrarily adjustable stiffness. In the following section, we will consider this point in more detail.

2. The cross section σ on the right-hand side of the VUU-equation. We use the measured isospin, energy, and angular dependence of the free-NN cross sections as input.

This clearly shows the split-up of the interactions into a potential-like term (left-hand side) and a hard-core-collision term (right-hand side).

8.4 The Nuclear Equation of State

In thermodynamics, the equation of state determines a relation between macroscopic variables allowing one variable (for example, the internal energy) to be written as a function of intensive and extensive thermodynamic quantities. The nuclear equation of state is usually given in such a way that the energy per baryon of infinite nuclear matter is written as a function of temperature T and density ϱ,

$$W_{\text{tot}} = W(\varrho, T, \text{spin, isospin, } \Delta\boldsymbol{p}, \ldots,). \tag{8.6}$$

In practice, the dependence on spin, isospin, and momenta is neglected, motivated by the fact that infinite nuclear matter is a homogeneous and spin- and isospin-saturated system. The total energy is split up into a ground-state energy W_0, a compressional part W_{C} depending only on the density, and the thermal part W_{T}:

$$W_{\text{tot}} = W_{\text{T}}(\varrho, T) + W_{\text{C}}(\varrho, T=0) + W_0(\varrho = \varrho_0, T=0). \tag{8.7}$$

Since the thermal energy W_{T} should vanish for $T = 0$, we can define the compression energy W_{C} in the following way:

$$W_{\text{C}}(\varrho) = W_{\text{tot}}(\varrho, T=0) - W_{\text{tot}}(\varrho = \varrho_0, T=0). \tag{8.8}$$

Different ansatzes for the compressional energy W_{C} may be used. In hydrodynamics, a quadratic ansatz stemming from the extended liquid-drop model was often used [8.2]. In the Skyrme [8.6] ansatz, the potential is given by a sum of two- and three-body interactions,

$$V = \sum_{i<j} V_{ij}^{(2)} + \sum_{i<j<k} V_{ijk}^{(3)}, \tag{8.9}$$

where

$$\begin{aligned} V_{ij}^{(2)} &= t_0(1 + x_0 P_\sigma)\delta(\boldsymbol{r}_i - \boldsymbol{r}_j) + \frac{1}{2}t_1\left[(k^+)^2\delta(\boldsymbol{r}_i - \boldsymbol{r}_j) + \delta(\boldsymbol{r}_i - \boldsymbol{r}_j)k^2\right] \\ &\quad + t_2\boldsymbol{k}^+\delta(\boldsymbol{r}_i - \boldsymbol{r}_j)\boldsymbol{k} + iW_0\boldsymbol{k}^+\delta(\boldsymbol{r}_i - \boldsymbol{r}_j)\sigma \cdot \boldsymbol{k}, \end{aligned} \tag{8.10}$$

with $\boldsymbol{k}$ the operator of the relative momentum $\boldsymbol{k} = \frac{1}{2i}(\nabla_i - \nabla_j)$ and P_σ the operator of the spin exchange. The three-body term is

$$V^{(3)}(\boldsymbol{r}_1, \boldsymbol{r}_2, \boldsymbol{r}_3) = t_3\delta(\boldsymbol{r}_1 - \boldsymbol{r}_2)\delta(\boldsymbol{r}_1 - \boldsymbol{r}_3). \tag{8.11}$$

For determining the total energy, simplifying assumptions like spin and isospin saturation and neglection of the spin interactions are made.

The ground-state energy is derived using the Hartree–Fock method. The wave function of the total system Ψ is a Slater determinant of single-particle wave functions:

$$E_0 = \left\langle \Psi | \hat{T} + \hat{V}^{(2)} + \hat{V}^{(3)} | \Psi \right\rangle = \int H(r) \mathrm{d}^3 r. \tag{8.12}$$

We use the abbreviations τ for the energy density and ϱ for the baryon density:

$$\varrho = \sum_{i=1}^{A} |\psi_i(r)|^2 = \frac{2}{3} \frac{k_{\mathrm{F}}^3}{\pi^2}; \qquad \tau = \sum_{i=1}^{A} |\nabla \psi_i(r)|^2 = \frac{3}{5} \frac{k_{\mathrm{F}}^2}{\varrho}. \tag{8.13}$$

After lengthy calculations, $H(r)$ becomes

$$H(r) = \hbar^2 \frac{\tau}{2m} + \frac{3}{8} t_0 \varrho^2 + \frac{1}{16} t_3 \varrho^3 + \frac{1}{16}(3t_1 + 5t_2)\rho\tau + \frac{1}{64}(9t_1 - 5t_2)(\nabla\rho)^2. \tag{8.14}$$

The term with $(\nabla\varrho)^2$ is suppressed in infinite homogeneous nuclear matter. In finite nuclei, these terms may have importance on the surface of the nucleus. On expanding these terms in first order, one gains a term proportional to ϱ^2 [which yields a slight reduction of the term containing t_0 in $H(r)$] plus sums over Yukawa-type two-body potentials. The term proportional to $\varrho\tau$ is also suppressed because of neglection of the relative momenta. Terms corresponding to the Yukawa type and to momentum dependence of the particles are implemented, for instance, in the quantum-molecular-dynamics (QMD) model [8.11], but are not considered in VUU.

Thus, only the first three terms on the right-hand side remain. The first term can be rewritten as $3/5 E_{\mathrm{F}} \varrho$ because of the definition of τ. If we now divide $H(r)$ by ϱ, we get the energy per baryon:

$$\frac{E}{A} = \frac{H}{\rho} = \frac{3}{5} E_{\mathrm{F}} + \frac{3}{8} t_0 \rho + \frac{1}{16} t_3 \rho^2. \tag{8.15}$$

The potential is the derivative of H with respect to ϱ:

$$U = \frac{\partial H}{\partial \rho} = \frac{3}{4} t_0 \rho + \frac{3}{16} \rho^2 = \alpha\rho + \beta\rho^2. \tag{8.16}$$

As a generalization, one may write

$$U = \alpha\rho + \beta\rho^\gamma \tag{8.17}$$

as in the *modified Skyrme forces.* In order to fix the three parameters α, β, and γ, one needs three conditions:

1. The ground-state energy E/A must take the correct value for nuclear matter $E_{\mathrm{B}} = -16$ MeV.

2. The ground state at $\varrho = \varrho_0$ has to be a minimum in the equation of state, that is, the first derivative of $E_C(\varrho, T = 0)$ with respect to ϱ should vanish. This is equivalent to the condition that the pressure be zero,

$$P = \left.\frac{\partial E}{\partial V}\right|_{N_B} \rho^2 \frac{\partial}{\partial \rho}\left(\frac{E}{A}\right) = 0. \tag{8.18}$$

3. The compression modulus κ should be some hundred MeV. The magnitude of κ can be obtained by measuring the change of the radius of nuclei in giant monopole resonances:

$$\kappa = R^2 \left[\frac{\partial^2}{\partial R^2}\left(\frac{E}{A}\right)\right]_{A,T=\text{const}} = 9\rho^2 \frac{\partial^2}{\partial \rho^2}\left(\frac{E}{A}\right). \tag{8.19}$$

In this way, several parametrizations of the nuclear equation of state or "different equations of state" can be obtained. In the following, we will often abbreviate the term "equation of state" by the initial letters "eos."[1] As an example, we write two parametrizations commonly used in VUU:

- A repulsive potential of high compressibility ($\kappa = 380$ MeV), a so-called "hard eos,"

$$U = -124 \text{ MeV } \frac{\rho}{\rho_0} + 70.5 \text{ MeV } \left(\frac{\rho}{\rho_0}\right)^2. \tag{8.20}$$

- A less repulsive "soft eos" with $\kappa = 200$ MeV,

$$U = -356 \text{ MeV } \frac{\rho}{\rho_0} + 303 \text{ MeV } \left(\frac{\rho}{\rho_0}\right)^{\frac{7}{6}}. \tag{8.21}$$

8.5 Numerical Realization

Integration of the VUU equation is done by propagating an ensemble of N events calculated in parallel, each with A point-like nucleons simulating the function f, with special starting conditions for f at $t = 0$. The Monte–Carlo approximation for the distribution function f is obtained as

$$f = \frac{1}{N} \sum_{\text{events}} \sum_{i=1}^{A} \delta(\boldsymbol{r}_i - \boldsymbol{r})\delta(\boldsymbol{p}_i - \boldsymbol{p}). \tag{8.22}$$

The distribution function at a phase-space point is approximated by drawing a sphere in phase space, that is, by counting all particles of all events within a phase-space cell. The density, for example, is then given by

$$\varrho(\boldsymbol{r}) = \int f \mathrm{d}^3 p = \frac{3}{N\, 4\pi R_\varrho^3} \sum_{\text{events}} \sum_{i=1}^{A} \Theta(R_\varrho - |\boldsymbol{r}_i - \boldsymbol{r}|), \tag{8.23}$$

[1] "Eos" is the Greek goddess of the dawn.

where the radius of the sphere R_ϱ (in the program the parameter FRAD) should be neither too large, to avoid non-local interactions, nor too small, to prevent the system from being influenced excessively by fluctuations in the distribution of the particles. In the present program with $N = 15$ parallel events, R_ϱ is set to 2 fm.

For the starting configuration, we use the Fermi-gas ansatz. All particles of one nucleus are randomly distributed inside a sphere in coordinate and momentum space:

$$(\boldsymbol{r}_i - \boldsymbol{r}_{\mathrm{CM}})^2 \leq r_0^2 A^{\frac{2}{3}}, \quad (\boldsymbol{p}_i - \boldsymbol{p}_{\mathrm{CM}})^2 \leq \boldsymbol{p}_{\mathrm{F}}^2 \quad \text{for all } i. \tag{8.24}$$

The particles are then propagated according to classical equations of motion. The left-hand side of the VUU equation is solved with the forces on the particles calculated from the gradient of the density-dependent potential,

$$\boldsymbol{F}(\boldsymbol{r}) = -\boldsymbol{\nabla} U\left[\varrho(\boldsymbol{r})\right] = -\frac{\mathrm{d}U\left[\varrho(\boldsymbol{r})\right]}{\mathrm{d}\varrho} \cdot \boldsymbol{\nabla}\varrho(\boldsymbol{r}). \tag{8.25}$$

As already stated, in our simulation model the density and its gradient are determined by drawing a sphere around each particle. This corresponds to a comoving Lagrangian method of determining the total density. Another possibility, used in the so-called "BUU-model" [8.8], is to calculate the density on a fixed grid and to take the density from the box in which the particle resides, which corresponds to a fixed Eulerian method. Both methods yield similar results [8.12]. The Lagrangian method works more precisely for dilute systems, while the Eulerian method works faster in dense systems.

For the solution of the right-hand side, stochastic scattering is used, that is, a pair of particles collides if their minimum distance d fulfills

$$d \leq d_0 = \sqrt{\frac{\sigma_{\mathrm{tot}}}{\pi}}, \qquad \sigma_{\mathrm{tot}} = \sigma(\sqrt{s},\ \mathrm{type}). \tag{8.26}$$

If the particles scatter, their new momenta are randomly chosen using conservation of total energy and momentum. Afterwards, the Pauli principle is checked, that is, the Uehling–Uhlenbeck blocking factors $(1 - f_1')(1 - f_2')$ are calculated; these factors are regarded as probabilities that the respective final state phase cells are not occupied and the collision probability is reduced accordingly. For example, if $f_i' = 0.6$ in the final state, the final state of collision partner i will be allowed in 40% of all cases. Since the Pauli blocking has to be fulfilled for both particles, only about 16% of the collisions would be allowed if $f_1' = f_2' = 0.6$. In practice, two random numbers a_1 and a_2 are taken, drawn from a uniform distribution in the interval [0,1]. The collision is only allowed if the conditions

$$a_i \leq 1 - f_i', \qquad (0 \leq a_i \leq 1) \qquad i = 1, 2 \tag{8.27}$$

are fulfilled.

In the VUU model, these collisions may only occur between particles of the same event. Other implementations of the same equation, for example, the so-called Landau–Vlasov (LV) equation [8.9], allow collisions between particles in different events. This is, the same as using only one event and representing each nucleon by N test particles, where every test particle may collide with any other test particle. In this case, the cross sections have to be scaled to $\sigma_{\mathrm{LV}} = \sigma/N$. The second method (LV) has the advantage of achieving a higher localization of the simulation and of yielding a better approximation to the Boltzmann equation, whereas it has the disadvantage of a much longer calculation time for large systems.

The QMD model [8.11] corresponds to using Gaussian distributions for the test particles in configuration and momentum space. It allows a treatment of many-body effects: the N-body classical correlations are treated just as in the classical molecular-dynamics model [8.13], while taking into account the most important quantum effects, namely the Pauli principle, Fermi motion, particle creation and annihilation, and smooth wave functions.

In the treatment of the collision term, the selection of elastic and inelastic branches has to be done. This is performed randomly once a collision occurs. The condition for having a collision is

$$d^2\pi \leq \sigma_{\mathrm{tot}} = \sigma_{\mathrm{el}} + \sum_{i=1}^{n_{\mathrm{in}}} \sigma_i . \tag{8.28}$$

A random number a, $0 \leq a \leq 1$, is used to select between the following branches:

$$\begin{aligned} &\text{elastic} && \text{if} \quad a \leq \frac{\sigma_{\mathrm{el}}}{\sigma_{\mathrm{tot}}}, \\ &\text{branch } j && \text{if} \quad \frac{\sigma_{\mathrm{el}} + \sum_{i=1}^{j-1} \sigma_i}{\sigma_{\mathrm{tot}}} < a \leq \frac{\sigma_{\mathrm{el}} + \sum_{i=1}^{j} \sigma_i}{\sigma_{\mathrm{tot}}}. \end{aligned} \tag{8.29}$$

In the present model, only the $\Delta(1232)$ resonance and its decay into nucleon and pion are included. The pion can also be absorbed by a nucleon forming a delta. Therefore, there are four different inelastic collision types:

1. $NN \rightarrow N\Delta$ (Δ production)
2. $N\Delta \rightarrow NN$ (Δ absorption)
3. $\Delta \rightarrow N\pi$ (Δ decay – π production)
4. $N\pi \rightarrow \Delta$ (π absorption).

Of course, the different isospin channels have to be taken into account with their Clebsch–Gordan coefficients; for example, for the delta decay we have

$$\begin{aligned} \Delta^{++} &\rightarrow 1 \times (p + \pi^+), \\ \Delta^{+} &\rightarrow \frac{2}{3} \times (p + \pi^0) + \frac{1}{3} \times (n + \pi^+), \end{aligned}$$

$$\begin{aligned} \Delta^0 &\rightarrow \frac{2}{3} \times (n + \pi^0) + \frac{1}{3} \times (p + \pi^-), \\ \Delta^- &\rightarrow 1 \times (n + \pi^-). \end{aligned} \tag{8.30}$$

The selection of these branches is also done stochastically, as described for the selection of the elastic and inelastic branches.

8.6 The Basic Structure of the Program

The program VUU is contained in the file `VUU.FOR`. The basic parameters, COMMON-blocks, and the cross-section parameters given in the files `VUUPRM`, `VUUDIM`, and `VUUSIG` should be included in the headers of the subroutines. The basic structure of the program is as follows.

First, the parameters of the simulation are input, for example, charges IZ1, IZ2 and masses N1, N2 of projectile and target, incident energy ELMEV, impact parameter DA(11), time-step width DA(12), number of time steps NDT, as well as internal parameters such as the selection of the eos for subroutine `SETEQS` [DA(16)] or the cross-section factor DA(8) in entry `CR1LBL`.

Then the selected number KQT of ensemble runs, each with NSIMUL $(=15)^2$ parallel events is calculated:

- Each event is initialized according to the Fermi-gas ansatz. The coordinates and momenta are randomly chosen within a sphere in coordinate and momentum space. The reference frame is the so-called "equal-speed system," where the velocities of projectile and target particles have the same absolute values.
- The system is propagated NDT time steps of size DT=DA(12). Each time step is treated in the same way:
  - At the beginning of each time step, the particle densities, the mean field potential, and the forces are calculated in subroutine `DENS`.
  - All possible collisions are checked in function `HIT`, their time of minimum distance is determined in subroutine `IMPACT`, and the energy-dependent cross section is calculated in subroutine `CROSS1`. All collisions that may occur are stored in a list in subroutine `COLOAD`. After each collision, the collision list will be updated in subroutine `CUPDAT` (and subroutine `DUPDAT` for the decays). The following will be done until no more collisions occur in the time step considered:
    - * Find the next collision in subroutine `FNDNXT`.
    - * Propagate to the time of the next collision in subroutine `TRSPRT`.
    - * Save the configuration of the collision pair in subroutine `SVEIFO`.

[2]This value is set via a PARAMETER declaration in `VUUPRM`. It should be changed only if necessary.

  - ∗ Perform the collision in subroutine SCATTR, or the delta decay in subroutine DELDEC, or a pion absorption in subroutine ABSORB. Generate new momenta.
  - ∗ Check the Pauli blocking in subroutine PAULI.
  - ∗ If the collision is Pauli blocked, restore the two-particle configuration in entry RSTIFO; else update the collision list in subroutines CUPDAT and DUPDAT.

  Transport to the end of the time step if no further collision occurs.
- If the number K of the time step is an integer multiple of NPRINT, print out the coordinates X, Y, Z (in fm), the momenta PX, PY, PZ (in GeV/c), the mass EM (GeV/c^2), the particle type ITYP, the number of the last collision partner LASTCL, and the collision counter NCLCNT of each particle on unit 7.

When sufficient statistics have been generated (for heavy systems at least 2–3 *runs* of 15 events each, for light systems 6–10 runs), the output files can be analyzed. As a starting point, the program VUUANL.FOR can be used (and further modified) for input and analysis.

8.7 Running the Program

8.7.1 Compilation and Linking

The main program is VUU.FOR, and it uses the INCLUDE files VUUPRM, VUUDIM, and VUUSIG. The file VUUANL.FOR contains a separate program to read the output. For the generation of random numbers between 0 and 1, we use the function DRAND48, while SRAND48 serves to initialize the seed of the randomizer. Both are UNIX routines and were used on the IBM RISC/6000 work station, on which the sample results were generated. The calls are done via subroutine RANF with entry SEED, which for convenience is placed at the end of the code. They may be replaced by any of the many available programs for generating uniformly distributed random numbers. The program as distributed uses unit 5 as standard input file, unit 6 as standard output for logging the run, and unit 7 for the output of the configuration. The logical record length of the file allocated to unit 7 should be at least 80 (this applies to IBM MVS systems mainly).

8.7.2 Input

The *first* line of the input contains the header of the simulation, that is, four characters characterizing the simulation. If your system allows dynamical allocation of files, you could use the header for the selection of the file name. The header **"TEST"** has a special meaning: in this case, the system will do a test run with alpha on alpha, and no further input lines are needed.

For the other input lines, reference is often made to an array "DA." This array of 22 elements is simply used for collecting the large number of parameters in a manageable block and allows modifying selected values through the input (see below). Below, the corresponding index in DA will be given for each case.

The *second* input line contains the mass and charge numbers of the projectile and of the target N1, IZ1, N2, and IZ2. The values of N1 and IZ1 will be stored in DA(1) and DA(2), and the values of N2 and IZ2 in DA(6) and DA(7).

The *third* input line contains the total number of time steps NDT to be run, the number of time steps between successive outputs of the configuration NPRINT, the number of runs KQT in the ensemble, and the kinetic energy of the incoming projectile in the LAB-system, ELMEV. The number NDT of time steps together with the width DA(12) (default = 0.25 fm/c) determines the total reaction time, for example, NDT = 200 with default parameter yields a time of 50 fm/c. The reaction time should be larger than the sum of projectile and target radii divided by the velocity β of projectile or target in the equal-speed system.

Finally, there is an input cycle for the DA-block. Each input line consists of the index of the element followed by the value. The indices 1, 2, 5, 6, 7, and 10 should not be used since they have already been defined in previous input lines. The input cycle is closed by using the index 0. When working in dialogue mode, the index −1 can be used to get a short description of the elements. We will first give an example before we explain the elements in detail.

Input	**Interpretation**
NENB	Header, in this case indicating neon + niobium
20 10 93 41	$A_{\mathrm{P}} = 20, Z_{\mathrm{P}} = 10, A_{\mathrm{T}} = 93, Z_{\mathrm{T}} = 41$
200 100 2 800.0	200 time steps, output every 100, 2 runs, 800 MeV per nucleon
−1 0.0	prints DA-shortlist on screen
11 3.0	sets DA(11) (impact parameter) to 3.0 (fm).
0 0.0	end of input.

In this example, the collision runs for 50 fm/c, and the output occurs after 25 and 50 fm/c.

Elements of the parameter block DA:

DA(1) is the projectile mass number. It is set by the second input line.

DA(2) is the projectile charge number, set by the second input line.

DA(3) is the radius r_0 in fm for the determination of the projectile and target radii via $R = r_0 A^{1/3}$. The default is 1.12 fm.

DA(4) determines the initial position of the projectile center-of-mass along the z-axis in units of its radius via $z_{\mathrm{P}} = \mathrm{DA}(4) \times R_{\mathrm{P}}$. The default value is 1.0. Since the projectile is assigned a negative momentum in the z-direction, this displacement should be positive.

DA(5) is the initial momentum per nucleon of the projectile in GeV, determined from ELMEV. It has a negative value.

DA(6) is the target mass number set by the second input line.

DA(7) is the target charge number set by the second input line.

DA(8) is the cross-section factor for determining the effective cross section in terms of the free one, $\sigma_{\text{eff}} = \text{DA}(8) \times \sigma_{\text{free}}$. Its default value is 1.0.

DA(9) is [like DA(4) for the projectile] the factor for the initial z-component of the target center-of-mass. It should have a negative value. The default is -1.0.

DA(10) is the momentum per nucleon of the target. Since the reference frame is the equal-speed system, the value of DA(10) is the absolute value of DA(5).

DA(11) is the impact parameter in fm. The default is 0.0 fm. If DA(11)<0, a random impact parameter (with square weight) between 0 and $-$DA(11) is selected.

DA(12) is the length of one time step in fm/c. The default is 0.25 fm/c.

DA(13) is a factor changing the delta decay width of 120 MeV. The default is 1.0.

DA(14) is used to set some switches. The meaning of the switches and how to set them will be explained after this list of the DA-elements. The default is 0.0 (no switch set).

DA(15) is a factor allowing one to change the Fermi momentum, the maximum momentum in the initialization. $P_{\text{F}}^* = \text{DA}(15) \times P_{\text{F}}$, where P_{F} depends on the ground-state density and thus on DA(3). With the default value of DA(3), we have $P_{\text{F}} \approx 268$ MeV. The default value of DA (15) is 1.0.

DA(16) determines the equation of state for the potential U. For DA(16)>0, DA(16) is the incompressibility of the eos. For DA(16)<0, several standard parameter sets can be selected (see below). For DA(16)=0, no mean-field potential is implemented (cascade mode). The default is -2 (parameter set for a hard eos).

DA(17) controls the cutoff for the preselection of the collisions. This preselection prevents particles far away from each other being tested for collisions and helps to save time. The default is 1.0, a reduction saves time, but should not be done at too high energies.

DA(18) is the minimum energy cutoff in GeV if no Pauli blocking is included. To avoid too soft collisions, only collisions with a relative energy higher than the minimum energy are allowed. If Pauli blocking is included, this minimum energy is set to 1 MeV, otherwise, the value is set to DA(18). The default is 0.05 GeV.

DA(19) is the size of a confining box. If cascade mode is selected [DA(16)=0.0] and DA(19)>0, the system is confined in a cubic box with side length DA(19) in fm. This is only interesting when considering the thermal models, not for normal simulation of collisions; therefore, the default is 0.0.

DA(20) is the seed for initializing the randomizer. For improving the statistics, different runs should be started with different seeds.

DA(21) is the chemical potential in MeV of a Fermi gas for simulating thermal models. This mode can only be run if only one "nucleus" is set up and the target nucleus mass N2 is set to 0 and DA(22)>0. For normal simulation of heavy-ion collisions, this mode is not used. Therefore, the default is 0.0.

DA(22) is the temperature of a Fermi gas in MeV if a thermal model is applied. The condition for this mode is the same as in the description of DA(21): N2=0 and DA(22)>0. The default is 0.0.

Switches: As mentioned, DA(14) controls various aspects of the program as a collection of 8 switches. Each switch is set by adding a power of two to DA(14); in detail, the correspondence is

Switch No.	1	2	3	4	5	6	7	8
add to DA(14)	1	2	4	8	16	32	64	128

For example, to set switches 3, 4, and 6, you should set DA(1)= 4+8+32 = 44. The meaning of the switches is as follows:

No. 1: if set, forces the delta-decay width to be constant 120 MeV, otherwise the delta-decay width will be mass dependent.

No. 2: if set, gives the deltas an infinite lifetime, otherwise, the lifetime will be randomly chosen within the decay width.

No. 3: if set, causes an output of the configuration after the first time step. This allows recording the initialization configuration.

No. 4: if set, disables the Pauli blocker. To avoid too soft collisions, an energy cutoff of DA(18) (in GeV) is set.

No. 5: if set, fixes the delta mass to 1232 MeV, otherwise, the mass is selected randomly within the allowed range for delta masses.

No. 6: if set, freezes the Fermi momentum up to the time of the first collision. This is only possible if the cascade mode is selected [DA(16)=0.0] and serves to keep the nuclei stable.

No. 7: if set, lets all deltas decay before the output of the final time step. Thus, the final distribution of all "stable" particles will be printed out.

No. 8: if set, reduces the collisions by allowing collisions between particles from the same initial nucleus only if at least one of them has collided with a particle from the other nucleus before. This enhances the stability of the nuclei.

Equations of state: if DA(16)<0 this parameter determines the equation of state as follows:

- DA(16) =−1, a "soft" eos with a compressibility of 200 MeV.
- DA(16) =−2, a "hard" eos with a compressibility of 380 MeV.
- DA(16) =−3, a "supersoft" eos with a constant potential for $\varrho \geq \varrho_0$.
- DA(16) =−4, a "very soft" eos with a compressibility of 121 MeV.
- DA(16) =−5, an "intermediate" eos with a compressibility of 284 MeV.
- DA(16) =−6, a "very hard" eos with a compressibility of 456 MeV.

8.8 Input Examples

- A calculation of Nb + Nb at 400 MeV, b = 3 fm, using a hard eos. Time step size is 0.2 fm, 400 time steps total with output at 200 and 400.

Input line	Comment
NBNB	
93 41 93 41	Nb + Nb
400 200 3 400	400 time steps with 200 output interval, 3 runs at 400 MeV
11 3.0	b =3 fm
12 0.2	0.2 fm/c steps
16 −2.0	hard eos
20 139.	seed for randomizer
0 0.0	end of input

- The same system with a soft eos and cross section reduced by a factor of 0.7.

Input line	Comment
SOFT	
93 41 93 41	same system
400 100 2 400	same time step control
8 0.7	$\sigma = 0.7\sigma_{\text{free}}$
11 3.0	b =3 fm
12 0.2	0.2 fm/c time step size
16 −1.0	soft eos
20 139.	same seed, same initialization
0 0.0	end of input

- A cascade calculation of Ne + Pb with frozen Fermi distribution, no Pauli blocking, $b < 3$ fm, 800 MeV, impact parameter randomly chosen below 3 fm.

Input line	Comment
CASC	
20 10 207 82	Ne on Pb
300 200 4 800.	300 steps, output at 200 (and 300!), 4 runs at 800 MeV
11 −3.0	$b \leq 3$ fm, randomly chosen
12 0.2	time step 0.2 fm/c
14 40.	32+8 = switches 4 and 6
16 0.0	cascade
18 0.06	60 MeV min. energy cutoff
20 555.	seed
0 0.0	end of input

- Finally, Ca on Ca at $b = 2$ fm with an eos of 140 MeV compressibility and a Fermi momentum of about 190 to 200 MeV.

Input line	Comment
CACA	
40 20 40 20	Ca on Ca
400 400 8 650.	400 steps, output at end, 8 runs 650 MeV
11 2.0	$b = 2$ fm
12 0.2	time step of 0.2 fm/c
14 4.0	output after first time step also
15 0.75	0.75 $P_{\text{F}}^{\text{normal}}$
16 140.	compressibility of 140 MeV
20 111.	seed
0 0.0	end of input

8.9 Interpretation of the Output

To understand the arrangement of the output, recall from the discussion in Sect. 8.6 that each calculation consists of a number of *runs*, each in turn consisting of NSIMUL (predefined as 15) *events*. The difference is that the events are coupled in the determination of the local density, while different runs are used only to accumulate better statistics.

The output is correspondingly divided up into runs, and in each run the printouts for given times form separate output sequences. It starts with a header for each output sequence, whose first line contains a counter $99 - I$, for the Ith output, followed by the header of the VUU-input, the number of time steps, and the physical time the output corresponds to, and the number of runs. Then the DA-block follows in five lines and, finally, the first line is repeated to conclude the header. After the last header block a conclusion line starting with 99-I_{last}-2 is printed out. For the first input example, the headers would be the following (the format is not exact):

98NBNB200.VUU 0.3999999E+02 3	first outp.: 200 steps, 40 fm/c
DA(1)–DA(5)	first number= proj. mass
DA(6)–DA(10)	first number= targ. mass
DA(11)–DA(15)	first number= impact par.
DA(16)–DA(20)	first number= eos
DA(21)–DA(22)	
98NBNB200.VUU 0.3999999E+02 3	end of first header
97NBNB400.VUU 0.7999999E+02 3	2nd outp.: 400 steps, 80 fm/c
DA(1)–DA(5)	first number= proj. mass
DA(6)–DA(10)	first number= targ. mass
DA(11)–DA(15)	first number= impact par.
DA(16)–DA(20)	first number= eos
DA(21)–DA(22)	
97NBNB400.VUU 0.7999999E+02 3	end of second header
95NBNB400.VUU 0.7999999E+02 3	end of header block

The counter number thus characterizes the output blocks according to the associated physical time in the simulations. These now follow, and each output block starts with the counter number followed by the numbers 5 and 45 (for the ICOUNT-block of VUU) and the number of the event inside the ensemble. This sequence is repeated for each run. For the first input example, the sequence would be the following:

98 5 45 1
output block of the first event of the first run after 200 time steps
98 5 45 2
output block of the second event of the first run

...
98 5 45 15
output block of the 15th event of the first run
97 5 45 1
first event of the first run after 400 time steps
...
97 5 45 15
last event of the first run after 400 time steps
98 5 45 1
first event of the second run after 200 time steps
...
97 5 45 15
last event of the second run after 400 time steps
...

Each output block has the following structure:

counter 5 45 number of event
counter NT=number of baryons NPION=number of pions impact parameter
counter / number of collisions / of Pauli-blocked collisions / of delta decays
10 lines of a collision statistics block (explained below).

For each particle, a line follows containing (in that sequence) the coordinates X, Y, and Z in fm, the momentum components PX, PY, and PZ in GeV/c, the mass EM in GeV, the particle type ITYP, the number LASTCL of the last collision partner, and the collision counter NCLCNT for each particle, indicating how many collisions the particle has experienced during the simulation up to now. In total, there will be NT + NPION lines.

The type of particle is determined by ITYP through the table

ITYP	1	2	3	4	5	6	7	8	9
particle	n	p	Δ^-	Δ^0	Δ^+	Δ^{++}	π^-	π^0	π^+

The collision-statistics block contains information about the number of collisions of a given type ICLTYP. ICLTYP codes the types of the two collision partners, ITYP1 and ITYP2 using the formula ICLTYP = $\frac{1}{2}$IMAX(IMAX $-$ 1) + IMIN with IMAX = MAX(ITYP1,ITYP2) and IMIN = MIN(ITYP1, ITYP2). For example, ICLTYP = 1 denotes an nn-collision, 2 = pn, 3 = pp, 20 = $\Delta^+\Delta^{++}$, etc.

The first line of this block lists the number of collisions for ICLTYP=1 to 21 for the elastic channel, the following 3 lines contain inelastic channels for the same ICLTYP values, the fifth line contains all Pauli blocked (elastic and inelastic) collisions of the same ICLTYP values. The next 5 lines contain in sequence the elastic channels for ICLTYP=22 to 45, 3 inelastic, and the Pauli-blocked channels.

The output can also be analyzed using the program `VUUANL.FOR`.

8.10 Output Analysis with VUUANL

The program `VUUANL.FOR` gives an example of reading in the output file for an analyzing program. `VUUANL` is not a complete analysis program, but it may be a convenient basis for writing specialized programs of this type.

The program consists mainly of subroutine `NXTEV` for reading the next event with entry `OPFILE` for reading the header. In order to let surviving deltas of the output decay, subroutine `DELDEC` with correlated functions is included. The requirements for a random-number generator are as for `VUU`, and again the sample routine is placed at the end.

First the program reads in a seed ISEED for the randomizer, which is needed when calculating the delta decay. If the deltas should not decay, you may enter a zero here. Afterwards, the system asks whether the analysis should be done in the lab system (IFLLAB=1) or in the equal speed system (IFLLAB=0).

Then a loop starts asking for the unit number of the output file to be read and for the time step counter of the output. A unit number of 0 finishes the input. A counter number 98 reads the output files of the first output section, and 97 reads those of the second output section. 0 reads in the last output section and a negative number reads in all output sections. Here are two examples:

23	seed for randomizer	0	no delta decay
1	lab system	0	equal speed system
7	unit 7	9	unit 9
0	last output block	−1	all output blocks
8	unit 8	0	end of input
97	second output block		
0	end of input		

8.11 Final Comments on Using the Program

If your FORTRAN library does not include the bit-manipulation functions IAND(i1,i2) and IOR(i1,i2), a simulation of these functions is contained in the file `VUUAUX.FOR`.

The major deviation from the FORTRAN-77 standard is the use of INCLUDE statements. In other systems, the syntax of these may have to be changed, or the inclusion may have to be done manually.

The output on units 5, 6, and 7 is done without an explicit preceding OPEN statement. Most compilers accept this and, at least for unit 7, assign some predetermined file name, which will also be used by VUUANL for determining the input file. If this does not work on your system, OPEN statements may have to be added.

In case of memory constraints, one may save vector space by removing the collision count vector NCLCNT from the program. Further, if the subroutines PAULI, DENS, and TRSPRT from the file VUU.FOR are replaced by the corresponding subroutines in the file VUUAUX.FOR, the COMMON block NOINFB with the vectors DENSPP and XYZSQ can be removed from the program.

If only small systems (for example, Ca + Ca) will be calculated, one may drastically decrease the vector length by setting the maximal total particle number per event NPEVMA from 600 to a lower value. In this case, you should also replace the maximal number of pions NPIONI by a smaller number. Both variables are declared as parameters in file VUUDIM. For decreasing the values of both parameters, it should be kept in mind that NPEVMA−NPIONI should not be smaller than the total baryon number NT of projectile and target. Furthermore, for energies not higher than 1 GeV, NPIONI should be at least one-fifth of the total baryon number NT; for example, for a maximal system of Ca on Ca NPEVMA=100 and NPIONI=20 may be useful.

8.12 Test Output

To test the general validity of the code, a set of test input and output files is provided. Unfortunately, because of the random nature of the simulations, the results will usually only be reproduced on the average, and it is more meaningful to compare the output of the statistical analysis of VUUANL rather than the individual events generated by VUU with the sample output files.

The file names are as follows: VUU.INP contains the input to VUU, which was also given as the last example in Section 8.8, VUU.OUT the initial part of the output on FORTRAN unit number 6, and VUU.DAT that on unit number 7, which is used also as the principal file read by VUUANL. The interactive input to VUUANL is given in VUUANL.INP and the resulting output on unit 6 in VUUANL.OUT.

References

[8.1] For a review and further references, see: H. Stöcker and W. Greiner, Phys. Rep. **137**, 277 (1986)

[8.2] W. Scheid, H. Müller, and W. Greiner, Phys. Rev. Lett. **32**, 741 (1974); H. Stöcker, J.A. Maruhn, and W. Greiner, Z. Phys. A**290**, 297 (1979), Phys. Rev. Lett **44**, 725 (1980), Phys. Rev. Lett **47**, 1807 (1981), Phys. Rev. C**25**, 1873 (1982); G. Buchwald, G. Graebner, J. Theiss, J.A. Maruhn, W. Greiner, and H. Stöcker, Phys. Rev. Lett **52**, 1594 (1984)

[8.3] H.A. Gustafsson et al., Phys. Rev. Lett. **52**, 1592 (1984); D. Beavis et al., Phys. Rev. C**27**, 2443 (1983); H.G. Ritter et al., Nucl. Phys. A**447**, 3c (1985); K.G.R. Doss, H.S. Gustafsson, H.H. Gutbrod, K.H. Kampert, B. Kolb, H. Löhner, B. Ludewigt, A.M. Poskanzer, H.G. Ritter, H.R. Schmidt, H. Wieman, Phys. Rev. Lett. **57**, 302 (1986); H.H. Gutbrod, K.H. Kampert, B.W. Kolb, A.M. Poskanzer, H.G. Ritter, H.R. Schmidt, Phys. Lett. **216B**, 267 (1989); K.H. Kampert, J. Phys. G15, 691 (1989)

[8.4] C.Y. Wong and T. Welton, Phys. Lett B**49**, 243 (1974); A.A. Amsden, G.F. Bertsch, F.H.Harlow, and J.R. Nix, Phys. Lett. **35**, 905 (1975); H.G. Baumgardt, J. U. Schott, Y. Sakamoto, E. Schopper, H. Stöcker, J. Hofmann, W. Scheid, and W. Greiner, Z. Phys. A**273**, 359 (1975)

[8.5] Y. Yariv and Z.Fraenkel, Phys. Rev. C**20**, 2227 (1979), Phys. Rev. C**24**, 488 (1981); J. Cugnon, Phys. Rev C**22**, 1885 (1980)

[8.6] T. H. R. Skyrme, Phil. Mag. **1**, 1043 (1956), Nucl. Phys. **9**, 615 (1956)

[8.7] H. Kruse, B.V. Jacak, and H. Stöcker, Phys. Rev. Lett. **54**, 289 (1985); J.J. Molitoris and H. Stöcker, Phys. Rev. C**32**, 346 (1985), Phys. Lett. B**162**, 47 (1985); J.J. Molitoris, H. Stöcker, and B.L. Winer, Phys. Rev. C**36**, 220 (1987)

[8.8] G.F. Bertsch, H. Kruse, and S. Das Gupta, Phys. Rev. C**29**, 673 (1984); J. Aichelin and G.F. Bertsch, Phys. Rev. C**31**, 1730 (1985); G.F. Bertsch and S. Das Gupta, Phys. Rep. **160**, 189 (1988)

[8.9] C. Gregoire, B. Remaud, F. Sebille, L. Vinet, and Y. Raffray, Nucl. Phys. A**465**, 317 (1987)

[8.10] E. A. Uehling and G. E. Uhlenbeck, Phys. Rev. **43**, 552 (1933), Phys. Rev. **46**, 917 (1934)

[8.11] J. Aichelin and H. Stöcker, Phys. Lett. B**176**, 14 (1986); J. Aichelin, A. Rosenhauer, G. Peilert, H. Stöcker, and W. Greiner, Phys. Rev. Lett. **58**, 1926 (1987); G. Peilert et al., Phys. Rev. C**39**, 1402 (1989); C. Hartnack et al., Nucl. Phys. A**495**, 303c (1989); M. Berenguer et al., *Proceedings of the NATO Advanced Research Workshop on Nuclear Matter and Heavy Ion Collisions*, edited by H. Stöcker and W. Greiner, NATO ASI Series B 205 (Plenum, New York, 1990) 343; C. Hartnack, H. Stöcker, and W. Greiner, *Proceedings on the NATO Advanced Research Course on The Nuclear Equation of State*, edited by H. Stöcker and W. Greiner, NATO ASI B 216, Vol. A (Plenum, New York, 1990) 239

[8.12] J. Aichelin and H. Stöcker, Phys. Lett. **163B**, 59 (1985); J. Aichelin et al., Phys. Lett. B**214**, 34 (1989)

[8.13] A.R. Bodmer, and C.N. Panos, Phys. Rev. C**15** (1977) 1342, Phys. Rev C22, 1023 (1980); S.M. Kiselev and Y.E. Prokovskil, Sov. J. Nucl. Phys. **38**(1), 46 (1983); J.J. Molitoris, J.B. Hoffer, H. Kruse, and H. Stöcker, Phys. Rev. Lett **53**, 899 (1984)

9. The Friction Model for Deep-Inelastic and Fusion Reactions

D.H.E. Gross

9.1 Introduction

Perhaps the most exciting and surprising discovery of nuclear reactions of heavy ions at lower intermediate energies was the discovery of a new, unexpected reaction type in the early 1970s, the deep-inelastic collisions [9.1–3]. At bombarding energies below about 10 MeV/nucleon, colliding nuclei convert most of the surplus kinetic energy above their mutual Coulomb barrier into internal heating. Despite this violent excitation, both nuclei leave the region of the strong interaction nearly intact. Moreover, this reaction type exhausts a major part of the total reaction cross section. That is, the bulk of the accessible phase space at these energies is populated by these reactions.

In contrast to well-known reactions at lower energies where only a few quantum channels get involved, we have here more macroscopic processes, and proper many-body aspects of the collision become important. Statistical concepts like entropy, dissipation, and fluctuations can be applied. In the 1970s, these pictures were new in nuclear physics. Recently, however, there has been a further development in this direction. At somewhat higher energies, there seem to be phase transitions, critical fluctuations, and – perhaps – intermittency in nuclear reactions [9.4,5].

That these macroscopic pictures apply to a microscopic system like an atomic nucleus, where quantum effects still play an important role, presents a special challenge. The combination of these divergent concepts – quantum coherence and statistical noise – in a single dynamical process has not yet been solved, by a long way.

At present, the only tractable method is to formulate a *classical* dynamical theory that takes account of the new statistical concepts like friction, fluctuating force, diffusion, and dissipation. The transport coefficients entering such a theory may be calculated microscopically and quantum mechanically. This is certainly only the easier part of the problem, the quantum-stochastic dynamics are still missing.

In the following, we sketch the simplest version of such a dynamical theory, the phenomenological model of surface friction (SFM), which was first introduced in 1972 [9.6,7]. Here, we assume that the two colliding nuclei become excited but do not otherwise change during the collision. Only the relative distance and orientation $\{r, \varphi\}$ of the two centers of the (spherical) nuclei are treated explicitly. The coupling of the relative motion of the two nuclei to their intrinsic motion is done by frictional forces. The motion remains in the scattering plane. Only elastic, inelastic, and fusion (that is, trapping) reactions can be described in this reduced version.

The effective radial potential V_{eff} is the sum of the nuclear potential V_{nucl}, the Coulomb potential $V_{\text{coul}} = e^2 Z_P Z_T / r$, and the centrifugal potential $V_{\text{centr}} = L^2/2\mu r^2$:

$$V_{\text{eff}}(r) = V_{\text{nucl}}(r) + V_{\text{coul}}(r) + V_{\text{centr}}(r). \tag{9.1}$$

For a given impact parameter, $d = \lim_{t\to-\infty} L/m\dot{r}$, the relative distance of both nuclei follows a trajectory $[r(t), \varphi(t)]$, which is the solution of Newton's equations with friction $(K_r\dot{r}, K_\varphi r\dot{\varphi})$:

$$\begin{aligned} \frac{\mathrm{d}}{\mathrm{dt}}(\mu\dot{r}) - \mu r\dot{\varphi}^2 + \frac{\mathrm{d}}{\mathrm{dr}}[V_{\text{nucl}}(r) + V_{\text{coul}}(r)] + K_r(r)\dot{r} &= 0, \\ \frac{\mathrm{d}}{\mathrm{dt}}(\mu r^2\dot{\varphi}) + K_\varphi(r) r^2\dot{\varphi} &= 0. \end{aligned} \tag{9.2}$$

The effective nuclear potential is approximated by the single folding of the nucleon density $\rho_i(r)$ and the nucleon–nucleus potential $V_j(r)$ of the other collision partner,

$$V_{ij}(\mathbf{r}) = \int \rho_i(\mathbf{r} - \mathbf{r}') V_j(\mathbf{r}')\, \mathrm{d}^3\mathbf{r}', \tag{9.3}$$

where we take an average fit for $\rho_i(r)$ and for $V_j(r)$ (see [9.7]):

$$\begin{aligned} \rho(r) &= \rho_0/\{1 + \exp[(r - R_D)/a_D]\}, \\ R_D &= 1.12A^{1/3} - 0.86A^{-1/3}\ \text{fm}, \quad a_D = 0.54\ \text{fm}, \\ \int \rho \mathrm{d}^3 r &= A, \quad \to \rho_0 = 0.170\ \text{fm}^{-3}, \quad \text{for } A > 12, \\ V(r) &= V_0/\{1 + \exp[(r - R_P)/a_P]\}, \\ R_P &= 1.25A^{1/3}\ \text{fm}, \quad a_P = 0.65\ \text{fm}, \\ V_0 &= -50\ \text{MeV}. \end{aligned} \tag{9.4}$$

The potential $V_{ij}(r)$ is usually symmetrized:

$$V_{\text{NN}}(r) = \frac{1}{2}[V_{ij}(r) + V_{ji}(r)]. \tag{9.5}$$

It is, of course, clear that this extremely simplified parametrization cannot describe observed individual fluctuations of the nucleus–nucleus potential away from the average trends. Only general trends can be described. It is nevertheless remarkable how well this very simple ansatz works.

In his more detailed application of the surface-friction model, P. Fröbrich [9.8] improved this ansatz by letting the parameters of the single nucleon potential change at the magic numbers. He suggests the following values for the parameters of the nucleon–nucleus potential:

		V_0	a_P
p-shell nuclei	$A \leq 16$	−30 MeV	0.35 fm
sd-shell nuclei	$A \leq 40$	−40 MeV	0.45 fm
nuclei above the sd-shell	$A > 40$	−50 MeV	0.65 fm

In order to speed up the numerical work, we fit an analytical form to V_{NN}:

$$V_{\mathrm{nucl}}(r) = -\sum_{n=1}^{5} A_n (r-R)^{n-1} \ln\left[1+\exp\left(-\frac{r-R}{a}\right)\right], \tag{9.6}$$

with $R = r_0(A_1 + A_2)^{1/3}$. In the fit, $r_0 = 1.3$ fm and $a = 0.61$ fm are kept fixed. Only the five coefficients A_n are optimized. They turn out to vary by only a little over the entire periodic table.

Relatively little is known a priori about the friction. Only at the far tails outside of the region of about 15% overlap of the two nuclei can one estimate the radial friction force by using perturbation-theoretical arguments [9.7,9,10]. These arguments indicate that the tangential part of the friction should be considerably weaker than the radial part. A quite successful parametrization of the r-dependence of the friction is obtained by linking the friction form-factor to the nuclear potential by the relations

$$\begin{aligned} K_r(r) &= K_r^0 \left[\frac{\mathrm{d}V_{\mathrm{nucl}}(r)}{\mathrm{d}r}\right]^2, \\ K_\varphi(r) &= K_\varphi^0 \left[\frac{\mathrm{d}V_{\mathrm{nucl}}(r)}{\mathrm{d}r}\right]^2. \end{aligned} \tag{9.7}$$

These relations are similar to the pertubation expressions for the friction from the force–force correlation-function [9.7]. The best average parameters throughout the periodic table are [9.7]

$$\begin{aligned} K_r^0 &= 4 \times 10^{-23}\ \mathrm{s/MeV}, \\ K_\varphi^0 &= 0.01 \times 10^{-23}\ \mathrm{s/MeV}. \end{aligned} \tag{9.8}$$

In the program, we call $K_r^0 := V_{RAD}$ and $K_\varphi^0 := V_{TAN}$. If dynamical deformations are treated explicitly as in the refinement of the friction model discussed by P. Fröbrich, then the strength K_r^0 has to be reduced to 3.5×10^{-23} s/MeV.

It is astonishing how universally this simple empirical parametrization of the friction works throughout the periodic table. This is certainly one of the strongest arguments in favor of the surface friction model [9.8].

Deep-inelastic reactions only give information about the total energy dissipation along the whole trajectory. Fusion is sensitive to the ingoing part only. Much more information on the local features of the dissipation is obtained in deep-inelastic break-up and incomplete fusion reactions. Here differential conservative and frictional forces can "shear" the colliding nuclei into parts. Consequently, we extended the friction model to an explicit treatment of the classical n-body dynamics of the α-particle subclusters in the projectile [9.11]. The results are very encouraging and support the relatively strongly rising friction at the nuclear surface.

In the following, we restrict ourselves to an introduction to the technical details of the surface friction model in its simplest form.

9.2 Description of the Program

The program is divided into three steps. (The names of the steps of the progam are written in capital letters):

A. The numerical calculation of the FOLDING-potential, $V_{\mathrm{NN}}(r)$, (9.5).

B. the FIT of the results from (A) by the analytic form, $V_{\mathrm{nucl}}(r)$, (9.6).

C. the solution of Newton's equations (9.2) with the potential $V_{\mathrm{nucl}}(r)$ and the action of the frictional force $K_r(r)\ \dot{r}$ and $K_\varphi(r)\ r\dot{\varphi}$ to obtain the FRICTIONal trajectories.

Most of the input and output of the programs goes into files with prescribed names; changing these can be done easily by changing the OPEN statements in the source files.

9.2.1 FOLDING

The purpose of this step is the calculation of $V_{12}(r)$ from (9.3). The input is:

(a) Via subroutine DATIN, a table of the Gauss–Legendre reference points and weights is copied from file `INPUT.DAT`.

(b) After this, a list of the systems is taken from `INPUT.DAT`. If more than one system should be calculated, lines with the relevant parameters are simply written one after the other:

APOT	=	A_{target};	ADICHT	=	$A_{projectile}$,
ZPOT	=	Z_{target};	ZDICHT	=	$Z_{projectile}$,
ENERLS	=	bombarding energy of the projectile (in MeV),			
LSTART, LEND	=	lowest and highest angular momentum (in $\hbar$).			

The method uses a transformation of the unbound two-dimensional integral in (9.3) by the substitutions

$$\begin{aligned}
r_1 &= |\mathbf{r}-\mathbf{r}'|, \\
r_2 &= |\mathbf{r}'|, \\
\mathrm{d}^3r' &= 2\pi y \mathrm{d}y \mathrm{d}z = 2\pi\frac{r_1 r_2}{r}\mathrm{d}r_1\mathrm{d}r_2, \\
u &= (r_1+r_2)/2, \\
v &= (r_1-r_2)/2, \\
\mathrm{d}^3r' &= 4\pi\frac{u^2-v^2}{r}\mathrm{d}u\ \mathrm{d}v. \qquad (9.9)
\end{aligned}$$

Then the folding integrals (9.3) become:

$$4\pi/r\int_{r/2}^{u_{\max}}\mathrm{d}u\int_{-r/2}^{+r/2}\mathrm{d}v,\ldots,$$

and a second one from

$$4\pi/r \int_{u_{\max}}^{2u_{\max}} du \int_{-r/2}^{+r/2} dv, \ldots,$$

where we choose

$$u_{\max} = r/2 + 2(R\textsc{pot} + R\textsc{dicht}). \tag{9.10}$$

It is checked whether the second integral is larger than 1% of the first; if this is so, the error message "precision not sufficient" is given.

The result is also printed out and stored in the file `FOLDING.DAT`.

9.2.2 FIT

Here, the symmetrized potential $V_{\mathrm{NN}}(r)$ from (9.5), obtained numerically in the previous section, is fitted by the analytical form $V_{\mathrm{nucl}}(r)$ (9.6) by a least-squares fit of the parameters $\{A_n\}, n = 1, \ldots, 5$.

The values obtained are written into the file `FIT.DAT` together with the values used for r_0 and a. A plot compares the original $V_{\mathrm{NN}}(r)$ and the fit $V_{\mathrm{nucl}}(r)$ linearly as well as logarithmically.

9.2.3 FRICTION

This part solves Newton's frictional equations (9.2) within the conservative potential $V_{\mathrm{nucl}}(r)$ from the previous step. For decreasing impact parameter (angular momentum L), the two-dimensional $\{r(t), \varphi(t)\}$ trajectories are calculated by using an integration routine with variable step size. The starting point is at $r = 2.5 \times R\textsc{frict}$, which is taken to be the sum of the radii at half density ($R\textsc{frict} = R\textsc{coul} + 2.7$ [fm], $R\textsc{coul} = \sum R\textsc{dicht}$). If, along the trajectory, the remaining local orbital angular momentum (called $ANG\textsc{loc}$ in the program)

$$ANG\textsc{loc}(t) = MR^2\dot{\varphi}(t) \tag{9.11}$$

is reduced to the sticking value of the angular momentum ($ANGM\textsc{st}$) or less, the coefficient of the tangential friction ($V\textsc{tan}$) is switched off. The sticking value of the orbital angular momentum is given by the condition that the angular velocity of relative motion $\dot{\varphi}$ becomes equal to the intrinsic angular velocity $\dot{\phi}$ of the two rotating nuclei. The latter is given by the loss of angular momentum, which is equal to the total angular momentum transferred from orbital motion to the internal motion:

$$\begin{aligned} L_{\mathrm{loss}}(t) &= L_{\mathrm{initial}} - ANG\textsc{loc}(t), \\ \dot{\phi} &= \frac{L_{\mathrm{loss}}}{\frac{2}{5}(M_1R_1^2 + M_2R_2^2)}. \end{aligned} \tag{9.12}$$

The trajectory is followed outwards to a distance of $2.6 \times R_{FRICT}$, where it is matched to a Coulomb trajectory. From this, the final energy and the angular momentum of relative motion and scattering angle are determined.

Next, the initial angular momentum L_{initial} is decreased by $\Delta L = 10\hbar$, and the calculation is repeated. When a trajectory is trapped for the first time, the initial angular momentum is reset to a value that is $30\hbar$ larger and the calculation is repeated with a reduced decrement of $L : \Delta L = 2\hbar$. The initial angular momentum of the first trapped trajectory defines the fusion cross section by the formula

$$SIG_{CF} = \pi\hbar^2 \times L^2_{initial}/(2 \times MASS \times c^2 \times ENER). \tag{9.13}$$

It is assumed here that there is no fusion window, that is, all lower impact parameters lead to fusion. This assumption, doubted for a long time, was never found to be violated.

At the end of this part of the program, the scattering function $[L, TTETA(L), E_{LOSS}(L)]$ for all L from the initial to the last one that does not fuse is printed. After this, the total fusion cross section (in mb) is also printed.

9.3 The Structure of the Final Printout File

First line: $A_{TAR}, A_{PRO}, Z_{TAR}, Z_{PRO}, E_{INITIAL}, L_{INITIAL}, L_{FINAL}$.

Second line: the parameters of the fitted potential: a_P, r_0, and the five parameters A_n.

Third line: the friction parameters.

Fourth line: the reduced mass, the coulomb parameter, angular momentum, and the cm-energy in units of (MeV,fm,10^{-23}s).

Following lines: These describe the trajectory: $Y(1) = r(t); Y(2) = \dot{r}(t); Y(3) = \varphi(t); Y(4) = \dot{\varphi}(t)$.

Final lines: The starting values for L, R, and φ follow, then the final values of L, R, and φ and the loss of orbital angular momentum and of energy. Next the distance of closest approach R_{MIN}, the orbital angular momentum L_{CLOSE}, the energy E_{CLOSE}, and the scattering angle $TETA$ are given.

This list is repeated for each successful trajectory, after which a table gives L, $TETA(L)$, $ENERGYLOSS(L)$, and the cross section for complete fusion (in mb).

9.4 Running the Code with the Sample Input

A sample input file INPUT.DAT is provided together with files that contain the corresponding output. To reproduce these, follow the following steps:

- Compile and link the FORTRAN source files FOLDING.FOR, FIT.FOR, and FRICT.FOR separately.
- Make sure the sample output files `FOLDING.DAT` and `FIT.DAT` are saved, because they will be overwritten by running the codes.
- Run FOLDING.FOR. This produces the output file `FOLDING.DAT`. The output on standard unit 6, which will normally appear on the terminal, is provided also in file `FOLDING.RES`.
- Run FIT.FOR, producing file `FIT.DAT` and standard output provided in file `FIT.RES`.
- Run FRICT.FOR, producing the final output on unit 6. The accompanying file `FRICT.RES` is shortened to show only the structure of the data and suffices for checking the validity of the output.

References

[9.1] R. Kaufmann and R. Wolfgang, Phys. Rev. **121**, 192 (1961)

[9.2] M. Lefort, C. Ngo, J. Peter, and B. Tamain, Nucl. Phys. A**197**, 485 (1972)

[9.3] A. G. Artukh, G. F. Gridnev, V. L. Mikheev, V. V. Volkov, and J. Wilczynski, Nucl. Phys. A**215**, 91 (1973)

[9.4] A. R. DeAngelis, D. H. E. Gross, and R. Heck, Nucl. Phys. A**537**, 606 (1992)

[9.5] D. H. E. Gross, A. R. DeAngelis, H. R. Jaqaman, Pan Jicai, and R. Heck, Phys. Rev. Lett. **68**, 146 (1992)

[9.6] R. Beck, D. H. E. Gross, and H. Kalinowski, in *International Conference in Nuclear Physics*, H. J. Mang et al., eds., (North-Holland, Amsterdam, 1973)

[9.7] D. H. E. Gross and H. Kalinowski, Phys. Rep. **45**, 177 (1978)

[9.8] P. Fröbrich, Phys. Rep. **116**, 337 (1984)

[9.9] S. Pal and D. H. E. Gross, Z. Physik A**329**, 349 (1988)

[9.10] L. Luo and D.H.E. Gross, Z. Phys. A**336**, 57 (1990)

[9.11] K. Möhring, T. Srokowski, and D. H. E. Gross, Nucl. Phys. A**533**, 333 (1991)

10. The Quark Model and the Nucleon–Nucleon Interaction

Amand Faessler and U. Straub

10.1 Introduction

There exists an almost general agreement that quantum chromodynamics (QCD) is the correct theory of the strong interaction. Thus, one should at least in principle be able to calculate the properties of hadrons and the interaction between hadrons from this theory. But it is well-known that owing to the large gluon–quark coupling constant and the non-linearity of the theory, one is not able to calculate the properties of hadrons and their interaction at low energies. At high momentum transfer, the running coupling constant $\alpha_S(q^2)$ assumes small values and perturbation theory can be applied. At high energies, one obtains with QCD good agreement with the experimental data. At low energies, the perturbative approach does not work. Thus, one tries two different methods: (i) One solves QCD numerically on a lattice. But with today's computer power one is not able to calculate reliable the properties of single mesons and baryons, not to mention the interaction between them. (ii) The second approach is the derivation of simpler models from QCD. We have today a wide variety of models that are QCD inspired and reproduce a large number of low-energy data for hadrons and their interaction.

The most successful of these models in terms of reproducing different observables is the non-relativistic quark model, which has been extensively applied to single baryons by Isgur and Karl [10.1,2]. In this model for the baryons, the constituent valence quarks are described by non-relativistic quark wave functions interacting by a one-gluon exchange and a quadratic confinement potential. A simple estimate naturally shows that a non-relativistic treatment is really not justified. But the model is extremely successful and can probably explain more observables than any other quark model. A possible reason for this is the fact that in a non-relativistic quark model, the center-of-mass motion can be separated off exactly, while this is not possible in a relativistic treatment. Gabriel Karl formulated this once in the following way: "It is better to treat the right degrees of freedom at a slightly wrong speed, than the wrong degrees of freedom at the right speed."

During the last eight years, our group in Tübingen has developed within this non-relativistic quark model a quantitative and reliable description of the nucleon–nucleon [10.3–6] and the nucleon–hyperon [10.7,8] interaction. Similar work has been done by the group of Yazaki and co-workers [10.9,10]. In these references, the group in Tokyo did not try to get a quantitative description, but instead were satisfied with explaining the short-range repulsion on the quark level. Recently, the Tübingen group has extended this model to the interaction of two hyperons [10.11]. We have calculated the binding energy of the so-called H-dibaryon [10.11] for the first time within a model that reproduces the nucleon–nucleon and the nucleon–hyperon data.

The first idea about the origin of the strong interaction between nucleons came from Yukawa in 1935. He postulated the existence of a medium-heavy particle (the meson) to explain the short-range nature of the nucleon–nucleon interaction. After the π-meson had been found in cosmic radiation, it was believed that now the interaction quantum of the strong force was known. But in the 1950s, it was shown that many different mesons need to be exchanged to explain the nature of the nucleon–nucleon interaction quantitatively. This approach of understanding the nucleon–nucleon interaction by exchange of mesons between the two nucleons culminated in the 1970s and 1980s in the Bonn potential [10.12–14] (see also Chapt. 1). These meson-exchange potentials are carefully fitted with about eight free parameters of coupling constants and cutoff masses to reproduce the two-nucleon data. But at short distances, the exchange of a meson between the center of one nucleon and the center of the second nucleon cannot be a correct description. At a distance of about 0.5 fm the two nucleons – each having a radius of the order of 1 fm – overlap almost completely. The antisymmetrization of the six quarks does not allow one to identify which three quarks belong to the first and which three quarks belong to the second nucleon. One obtains ten different contributions from the antisymmetrization, and in each of these ten different contributions, the centers of gravity of nucleon 1 and nucleon 2 are at different locations. Thus, for the short-range nature of the nucleon–nucleon interaction, the meson-exchange potentials can only be a phenomenological fit, but not a true description of the nature of the nucleon–nucleon interaction. This is, for example, also reflected in the fact that for the Bonn potential [10.12–14], one needs an ω-nucleon coupling constant $g_\omega^2/4\pi$, which is about three to five times as big as predicted by SU(3) flavor symmetry from the ρ-nucleon coupling constant. Thus, it seems that for the meson-exchange potentials one has to blow up the role of the ω-meson exchange out of all proportion to describe the short-range repulsion between two nucleons. In the quark model, one can understand the short-range repulsion [10.4] very naturally from the role of the orbital [42] symmetry of the six-quark wave functions. If one includes in addition to the quark picture ω-meson exchange, one obtains a best fit to the nucleon–nucleon data for an ω-nucleon coupling constant, which is within 20 % in agreement with flavor SU(3).

The computer code presented here is restricted to calculating the nucleon–nucleon phase shifts for the ^{1}S and ^{3}S partial waves. It includes, therefore, from the one-gluon exchange only the central part of the interaction. For larger distances, a linear or quadratic confinement potential is included. To obtain the medium and the long-range attraction, pion and σ-meson exchanges are included. The π-meson exchange can be calculated with the code in two ways: either as an exchange of the pions between the quarks or as an exchange between the nucleons including form factors. A third option allows one to include in addition ρ- and ω-exchanges on the baryon level.

The nucleon–nucleon scattering problem is solved with the resonating group equation including coupling to the $\Delta\Delta$ and hidden-color channels. The code calculates the norm and Hamilton kernels and computes from the solution of the resonating group equation the S-matrix and the nucleon–nucleon phase shifts.

10.2 The Resonating Group Method

In this section, we describe the resonating group method as a tool to solve the scattering problem of two interacting nucleons in the non-relativistic quark model. As interaction, we will take into account only central forces. This means that there is no coupling of different partial waves.

The general idea in this approach is to start from a six-quark wave function describing two interacting nucleons, each consisting of three quarks. In principle, this six-quark wave function should be determined from the Schrödinger equation taking an appropriate quark–quark interaction into account. But this six-particle problem is too complicated to be solved straightforwardly. Actually, we are mainly interested in obtaining the relative wave function, since this wave function describes the scattering of the two nucleons. On the other hand, the internal wave function determines the properties of a single nucleon. Also, the confinement of the quarks has to be reflected by the internal wave function. Finally, in this non-relativistic approach, the scattering problem is independent of the center-of-mass wave function. The idea to simplify the six-particle problem is therefore to write the six-quark wave function as a product of the internal wave functions of the two nucleons and the relative wave function and assume a reasonable ansatz for the internal wave function. It has been shown [10.1] that a successful ansatz for the internal wave function is a Gaussian. In the simplest approach, we choose this form of the internal wave function and do not allow the internal wave function to change during the collision. We are therefore left with the problem of determining only the relative wave function. This approach is realized in the present code.

A more refined way would be to allow for polarizations of the internal wave functions during the collision and determine this polarization dynamically together with the relative wave function from the Schrödinger equation [10.15].

Let us now put this simple approach in concrete terms. Using the standard definition of the internal Jacobi coordinates

$$\boldsymbol{\xi}_1 = \frac{\boldsymbol{r}_1 - \boldsymbol{r}_2}{2}, \qquad \boldsymbol{\xi}_2 = \boldsymbol{r}_3 - \frac{\boldsymbol{r}_1 + \boldsymbol{r}_2}{2},$$

$$\boldsymbol{\xi}_3 = \frac{\boldsymbol{r}_4 - \boldsymbol{r}_5}{2}, \qquad \boldsymbol{\xi}_4 = \boldsymbol{r}_6 - \frac{\boldsymbol{r}_4 + \boldsymbol{r}_5}{2},$$

$$\boldsymbol{R}_{AB} = \frac{\boldsymbol{r}_1 + \boldsymbol{r}_2 + \boldsymbol{r}_3}{3} - \frac{\boldsymbol{r}_4 + \boldsymbol{r}_5 + \boldsymbol{r}_6}{3}, \tag{10.1}$$

the six-quark wave function of the two interacting nucleons is written as [10.9]

$$\Psi_{6q}(\boldsymbol{\xi}^{\text{int}}, \boldsymbol{R}_{AB}) \equiv \mathcal{A} \left[\Phi^{\text{int}}(\boldsymbol{\xi}_1,\boldsymbol{\xi}_2)\, \Phi^{\text{int}}(\boldsymbol{\xi}_3,\boldsymbol{\xi}_4)\, \chi(\boldsymbol{R}_{AB})\right]_{ST}. \tag{10.2}$$

Φ^{int} is the internal wave function of the nucleons A, B. Although not explicitly written, these wave functions also contain the spin isospin and color parts of the nucleon wave function. Spin and isospin coupling is indicated by the bracket $[\,]_{ST}$. The Pauli principle is exactly taken into account by the antisymmetrizer $\mathcal{A}$ acting on the six quarks. Since by construction the wave function of a single nucleon is already antisymmetric [10.9,16], the antisymmetrizer of the six quarks has the simple form [10.16]

$$\mathcal{A} = \frac{1}{10}\left(1 - \sum_{\substack{k\in A\\ n\in B}} P_{kn}\right) \frac{1}{2}(1 - P_{14}P_{25}P_{36}). \tag{10.3}$$

The presence of the exchange operator $\frac{1}{2}(1 - P_{14}P_{25}P_{36})$ means that the two-nucleon wave function has to be antisymmetric even at the baryon level. This is expressed by the well-known selection rule $L + S + T = \text{odd}$.

For the internal wave function, we make the Gaussian ansatz (see also [10.9])

$$\begin{aligned}\Phi^{\text{int}}(\boldsymbol{\xi}_1,\boldsymbol{\xi}_2) &\equiv \left(\frac{1}{2\pi b^2}\right)^{3/4} \exp\left(-\frac{1}{4b^2}\boldsymbol{\xi}_1^{\,2}\right)\\ &\times \left(\frac{2}{3\pi b^2}\right)^{3/4} \exp\left(-\frac{1}{3b^2}\boldsymbol{\xi}_2^{\,2}\right).\end{aligned} \tag{10.4}$$

Since our aim is to describe the relative motion of the two clusters, we have to extract from the full six-quark Schrödinger equation the corresponding equation for the relative wave function $\chi(\boldsymbol{R}_{AB})$. This is attained by integrating over the internal degrees of freedom leading to the so-called Hill–Wheeler–Griffin equation:

$$\int \Phi^{\text{int}+}(\boldsymbol{\xi}_1,\boldsymbol{\xi}_2)\, \Phi^{\text{int}+}(\boldsymbol{\xi}_3,\boldsymbol{\xi}_4) \left(H - E\right) \Psi_{6q}(\boldsymbol{\xi}^{\text{int}}, \boldsymbol{R}_{AB})\, \mathrm{d}\boldsymbol{\xi}^{\text{int}} = 0. \tag{10.5}$$

An equivalent form of this equation is obtained by introducing the resonating group method (RGM) Hamiltonian kernels and norm kernels:

$$\begin{aligned}\mathcal{L}(\boldsymbol{R}', \boldsymbol{R}) &= \mathcal{H}(\boldsymbol{R}', \boldsymbol{R}) - E\, \mathcal{N}(\boldsymbol{R}', \boldsymbol{R})\\ &\equiv \int \Phi^{\text{int}+}(\boldsymbol{\xi}_1,\boldsymbol{\xi}_2)\, \Phi^{\text{int}+}(\boldsymbol{\xi}_3,\boldsymbol{\xi}_4)\, \delta(\boldsymbol{R}' - \boldsymbol{R}_{AB})\ (H - E)\\ &\quad\times \mathcal{A}[\Phi^{\text{int}}(\boldsymbol{\xi}_1,\boldsymbol{\xi}_2)\, \Phi^{\text{int}}(\boldsymbol{\xi}_3,\boldsymbol{\xi}_4)\, \delta(\boldsymbol{R} - \boldsymbol{R}_{AB})]\ \mathrm{d}\boldsymbol{\xi}^{\text{int}}\ \mathrm{d}\boldsymbol{R}_{AB}.\end{aligned} \tag{10.6}$$

Using these kernels, (10.5) can be written in the equivalent form

$$\int \mathcal{L}(\boldsymbol{R}', \boldsymbol{R})\, \chi(\boldsymbol{R})\, \mathrm{d}\boldsymbol{R} \;=\; 0. \tag{10.7}$$

This is the RGM equation.

A simple and fast way to solve this equation for a scattering state, which must satisfy the asymptotic boundary conditions of a scattering state, can be achieved by the following ansatz for the relative wave function (see [10.9,17]). First, we perform a partial-wave expansion:

$$\chi(\boldsymbol{R}_{AB}) \;=\; \sum_L \frac{1}{R_{AB}}\, u^L(R_{AB})\, Y_{LM}(\hat{\boldsymbol{R}}_{AB}), \tag{10.8}$$

and then we choose the generator coordinate ansatz for the radial part u^L:

$$u^L(R_{AB}) \;=\; \sum_{i=1}^{N} c_i\, \tilde{u}^L(R_{AB}, S_i), \tag{10.9}$$

$$\tilde{u}^L(R_{AB}, S_i) \;\equiv\; \begin{cases} p_i\, u^L(R_{AB}, S_i) & (R_{AB} \le R_c), \\ [h_L^{(-)}(k_{AB}R_{AB}) & \\ \quad -s_i\, h_L^{(+)}(k_{AB}R_{AB})]\, R_{AB} & (R_{AB} \ge R_c), \end{cases} \tag{10.10}$$

$$u^L(R_{AB}, S_i) \;\equiv\; 4\pi R_{AB} \left(\frac{3}{2\pi b^2}\right)^{3/4} \times \exp\left[-\frac{3}{4b^2}(R_{AB}^2 + S_i^2)\right]\; i_L\!\left(\frac{3}{2b^2} R_{AB} S_i\right). \tag{10.11}$$

S_i is called the generator coordinate. All the information about the relative wave function is now contained in the coefficients c_i. The condition $\sum_{i=1}^{N} c_i = 1$ is assumed for normalization. The complex parameters p_i and s_i are determined by equating the function and the derivative at $R = R_c$. This matching point R_c is chosen so that the interaction between the clusters can be neglected for $R_{AB} \ge R_c$. i_L is the Lth modified spherical Bessel function and $h_L^{(\pm)}$ is the Lth spherical Hankel function. For $L = 0$, one has $i_0(x) = \sinh x/x$ and $h_L^{(\pm)}(x) = \exp(\mathrm{i}x)/x$. The wave vector of the relative motion of the clusters is $k_{AB} = \sqrt{2\mu E_{\mathrm{rel}}}$, $\mu = 3m_{\mathrm{q}}/2$.

Obviously, for large separations $R > R_c$, the relative wave function satisfies the correct boundary condition of a scattering state with center-of-mass energy E_{rel}:

$$\chi^L(R_{AB}) = [h_L^{(-)}(k_{AB}R_{AB}) - S_L\, h_L^{(+)}(k_{AB}R_{AB})]\, R_{AB} \quad (R_{AB} \ge R_c). \tag{10.12}$$

For $R_{AB} \le R_c$, the basis functions $\tilde{u}^L(R_{AB}, S_i)$ are essentially Gaussians, which are peaked at S_i. This can be seen easily by projecting the function

$u^L(R_{AB}, S_i)/R_{AB}$, which describes the inner part of the relative wave function, on the partial wave L:

$$\chi^{\text{Gauss}}(\boldsymbol{R}_{AB}, \boldsymbol{S}_i) \equiv \left(\frac{3}{2\pi b^2}\right)^{3/4} \exp\left[-\frac{3}{4b^2}(\boldsymbol{R}_{AB} - \boldsymbol{S}_i)^2\right], \tag{10.13}$$

$$u^L(R_{AB}, S_i)/R_{AB} = \int \chi^{\text{Gauss}}(\boldsymbol{R}_{AB}, \boldsymbol{S}_i) \times Y^*_{LM}(\hat{\boldsymbol{S}}_i)\, Y_{LM}(\hat{\boldsymbol{R}}_{AB})\, \mathrm{d}\hat{\boldsymbol{S}}_i\, \mathrm{d}\hat{\boldsymbol{R}}_{AB}. \tag{10.14}$$

The scattering problem can now be solved by the Kohn–Hulthén–Kato variational method [10.18–20]. Defining the functional

$$J[\chi^L] \equiv S_L + \mathrm{i}\frac{\mu_{AB}}{k_{AB}} \int \chi^L(R')\, \mathcal{L}^L(R', R)\, \chi^L(R)\; R'\, R\, \mathrm{d}R'\, \mathrm{d}R, \tag{10.15}$$

one can show [10.17] that solving the RGM equation (10.7) together with the boundary conditions of a scattering state is equivalent to looking for an extremum of this functional. The RGM kernel $\mathcal{L}^L(R', R)$ projected on the partial wave L is defined as

$$\mathcal{L}^L(R', R) \equiv \int Y^*_{LM}(\hat{\boldsymbol{R}}')\; \mathcal{L}(\boldsymbol{R}', \boldsymbol{R})\; Y_{LM}(\hat{\boldsymbol{R}})\; \mathrm{d}\hat{\boldsymbol{R}}'\; \mathrm{d}\hat{\boldsymbol{R}}. \tag{10.16}$$

Inserting the ansatz (10.8–11) for the relative wave function and performing the variation with respect to the coefficients c_i, a set of linear equations for the coefficients c_i is derived [10.9,17]:

$$\sum_{j=1}^{N-1} \tilde{\mathcal{L}}_{ij}^{\;L}\; c_j = \mathcal{M}_i^{\;L}, \tag{10.17}$$

with

$$\tilde{\mathcal{L}}_{ij}^{\;L} = \tilde{\mathcal{K}}_{ij}^{\;L} - \tilde{\mathcal{K}}_{iN}^{\;L} - \tilde{\mathcal{K}}_{Nj}^{\;L} + \tilde{\mathcal{K}}_{NN}^{\;L}, \tag{10.18}$$

$$\mathcal{M}_i^{\;L} = \tilde{\mathcal{K}}_{NN}^{\;L} - \tilde{\mathcal{K}}_{iN}^{\;L}, \tag{10.19}$$

and

$$\tilde{\mathcal{K}}_{ij}^{\;L} \equiv \int \tilde{u}^L(R', S_i)\; \mathcal{L}^L(R', R)\; \tilde{u}^L(R, S_j)\; R'\; R\, \mathrm{d}R'\, \mathrm{d}R. \tag{10.20}$$

It should be noted that owing to the condition $\sum_{i=1}^{N} c_i = 1$, only $N-1$ of the coefficients are independent.

Having solved the linear problem (10.17), the S-matrix S_L in an elastic channel is given by

$$S_L = \sum_{i=1}^{N} c_i s_i, \tag{10.21}$$

which is the factor in front of the Hankel function h_L^+ of the wave function $u^L(R)$ at large R.

The kernels $\tilde{\mathcal{K}}_{ij}^L$ contain integrations over the wave function $\tilde{u}^L$, which has a different form in the inner and outer regions. A much simpler expression for this kernel can be obtained as shown in [10.17]:

$$\tilde{\mathcal{K}}_{ij}^L = p_i\, p_j\, \left[\mathcal{K}_{ij}^L - K_{ij}^{L\,(\mathrm{ex})}\right], \tag{10.22}$$

with

$$\begin{aligned}\mathcal{K}_{ij}^L = &\int \Phi_{6\mathrm{q}}(123,456;\boldsymbol{S}_i)Y_{LM}^*(\hat{\boldsymbol{S}}_i)\;(H-E)\\ &\times \mathcal{A}\;[\Phi_{6\mathrm{q}}(123,456;\boldsymbol{S}_j)]Y_{LM}(\hat{\boldsymbol{S}}_j)\;\prod_{k=1}^{6}\mathrm{d}\boldsymbol{r}_k\,\mathrm{d}\hat{\boldsymbol{S}}_i\,\mathrm{d}\hat{\boldsymbol{S}}_j,\end{aligned} \tag{10.23}$$

$$\begin{aligned}K_{ij}^{L\,(\mathrm{ex})} = &\int_{R_c}^{\infty} u^L(R,S_i)\left[-\frac{\hbar^2}{2\mu_{AB}}\frac{\mathrm{d}^2}{\mathrm{d}R^2}+\frac{\hbar^2 L(L+1)}{2\mu_{AB}R^2}-E_{\mathrm{rel}}\right]\\ &\times\; u^L(R,S_j)\,\mathrm{d}R.\end{aligned} \tag{10.24}$$

Here we have introduced the two-center shell-model wave function of the six quarks:

$$\Phi_{6\mathrm{q}}(123,456;\boldsymbol{S}_i) = \prod_{k=1}^{3}\Phi\left(\boldsymbol{r}_k,\frac{1}{2}\boldsymbol{S}_i\right)\;\prod_{n=4}^{6}\Phi\left(\boldsymbol{r}_n,-\frac{1}{2}\boldsymbol{S}_i\right), \tag{10.25}$$

$$\Phi(\boldsymbol{r},\boldsymbol{X}) \equiv \left(\frac{1}{\pi b^2}\right)^{3/4}\;\exp\left[-\frac{1}{2\,b^{\,2}}(\boldsymbol{r}-\boldsymbol{X})^2\right]. \tag{10.26}$$

The kernels $\mathcal{K}_{ij}$ thus contain only integrations over Gaussians, which can be performed completely analytically. Also, the action of the antisymmetrizer can be computed in the form (10.23) relatively easily. The remaining external kinetic-energy kernel $K_{ij}^{L\,(\mathrm{ex})}$ can be calculated by a simple numerical one-dimensional integration.

10.3 The Hamiltonian of the Six-Quark System

The strong interaction between the quarks is described by the non-relativistic Hamiltonian consisting of the kinetic-energy operator of the quarks and potentials on the quark level together with meson-exchange potentials either on the quark level or on the baryon level.

The various parts of the interaction have been studied in many publications. The short-range one-gluon exchange potential was introduced in [10.21]. The influence of this one-gluon exchange and the confinement together with the quark exchange has been studied in [10.3,4,9]. In these calculations, only short-range repulsion was obtained. Therefore, in [10.5,22], in addition to the gluonic potentials, meson-exchange potentials at the baryon level were included. Now a realistic description of the features of the nucleon–nucleon potential was possible, that is, the observed short-range repulsion and medium- and long-range attraction was obtained. Finally the pion exchange between the quarks was investigated in [10.6].

In the present code, we include the same interactions as in the references listed above. It should be noted, however, that the electromagnetic interaction, that is, the Coulomb force, is not taken into account. In addition, we restrict ourselves, as in the references above, to central forces. Hence, there is no tensor force and no spin–orbit force. This means that there will be no coupling of different partial waves. In the preceding chapter in the ansatz for the relative wave function, the restriction to central forces was already taken into account.

The Hamiltonian has the form

$$H = \sum_{k=1}^{6} \frac{\mathbf{p}_k^2}{2m_k} - K_{\mathrm{CM}} + \sum_{k<n} (V_{kn}^{\mathrm{OGE}} + V_{kn}^{\mathrm{conf}} + V_{kn}^{\pi}) + V_{\mathrm{NN}}^{\mathrm{OBEP}}. \tag{10.27}$$

The unphysical kinetic energy of the total center-of-mass motion K_{CM} is subtracted exactly.

As short-range potentials, we consider the one-gluon exchange and a phenomenological confinement potential:

$$V_{kn}^{\mathrm{OGE}} = \frac{\alpha_S}{4} \boldsymbol{\lambda}^c(k) \cdot \boldsymbol{\lambda}^c(n) \left\{ \frac{1}{r_{kn}} - \frac{\pi}{2} \delta(\boldsymbol{r}_{kn}) \left[\frac{2}{m_q^2} + \frac{4}{3} \frac{\boldsymbol{\sigma}(k) \cdot \boldsymbol{\sigma}(n)}{m_q^2} \right] \right\}, \tag{10.28}$$

$$V_{kn}^{\mathrm{conf}} = - \boldsymbol{\lambda}^c(k) \cdot \boldsymbol{\lambda}^c(n) \begin{cases} a^{\mathrm{lin}} \; r_{kn} & \text{(linear confinement)} \\ a^{\mathrm{quad}} \; r_{kn}^2 & \text{(quadratic confinement)} \end{cases}. \tag{10.29}$$

In our notation, the first part of the OGE potential is called the color–Coulomb potential, the second part the color–delta potential, and the third part the color–magnetic potential. As confinement potential, either a linear or a quadratic shape is used.

The long-range force due to pion exchange at the quark level has the following form:

$$\begin{aligned} V_{kn}^{\pi} = & \frac{1}{3} \frac{9}{25} \frac{g_{\mathrm{NN}\pi}^2}{4\pi} \frac{m_\pi^2}{4M_N^2} \frac{\Lambda^2}{\Lambda^2 - m_\pi^2} \mathrm{e}^{-m_\pi^2 b^2/3} \\ & \times m_\pi \left(\frac{\mathrm{e}^{-m_\pi r_{kn}}}{m_\pi r_{kn}} - \frac{\Lambda^3}{m_\pi^3} \frac{\mathrm{e}^{-\Lambda r_{kn}}}{\Lambda r_{kn}} \right) \boldsymbol{\sigma}_k \cdot \boldsymbol{\sigma}_n \; \boldsymbol{\tau}_k \cdot \boldsymbol{\tau}_n. \end{aligned} \tag{10.30}$$

If the pion exchange is included on the quark level, it must not, of course, be included at the baryon level and vice versa.

The meson exchange potentials at the baryon level consist of the π-, σ-, ω-, and ρ-meson exchange. Below we present only the form of the σ-meson-exchange potential, since this is the most important potential:

$$V^{\sigma}(r) = -\frac{g_\sigma^2}{4\pi} \frac{1}{2m_\sigma^2 R_\sigma^2 r} \begin{cases} 1 - \mathrm{e}^{-m_\sigma r} - \mathrm{e}^{-2m_\sigma R_\sigma} \sinh(m_\sigma r) & (r < 2R_\sigma) \\ [\cosh(2m_\sigma R_\sigma) - 1] \mathrm{e}^{-m_\sigma r} & (r > 2R_\sigma) \end{cases}. \tag{10.31}$$

The OBE potentials at the baryon level have to be included as a local potential in the RGM equation (10.7). In the corresponding interaction kernel at the quark level (10.23), this leads to a folding of the OBE potential with the relative-motion wave function.

Table 10.1. List of interaction diagrams and multiplicities

Interaction diagrams	Multiplicities
V_{12}	6
V_{36}	9
$V_{12}P_{36}$	2×9
$V_{36}P_{36}$	1×9
$V_{13}P_{36}$	4×9
$V_{16}P_{36}$	4×9
$V_{14}P_{36}$	4×9

The calculation of the interaction kernels (10.23) is more complicated for the potentials (10.28–30) at the quark level. At first glance, there are 15 pairs of interacting quarks, while the antisymmetrizer consists of another 10 permutation operators. Therefore, in principle, 150 diagrams have to be calculated. But the symmetry of the cluster model reduces the number of different diagrams greatly. Taking the antisymmetry of each three-quark cluster into account, we find that only seven topologically different diagrams remain: two different types of direct diagrams and five different types of exchange diagrams [10.9,16]. All of them have to be computed separately. A complete list of the non-equivalent diagrams is given in Table 10.1 together with the corresponding multiplicities.

The parameters of the Hamiltonian are adjusted to reproduce the mass splitting between the delta and the nucleon and the stability condition of the nucleon [10.3–6,22]:

$$\begin{aligned} M_\Delta - M_\mathrm{N} &= \frac{4}{3}\frac{\alpha_\mathrm{S}}{\sqrt{2\pi}\, m_\mathrm{q}^2\, b^3} + M_\Delta^\pi - M_\mathrm{N}^\pi, \\ \frac{\mathrm{d}}{\mathrm{d}b} M_\mathrm{N} &= 0 \quad \text{(stability condition).} \end{aligned} \tag{10.32}$$

M_Δ^π and M_N^π are the contributions of the pion clouds to the delta and the nucleon mass, respectively. The determination of the parameters proceeds as follows. One first chooses reasonable values for the size parameter b and the constituent quark mass m_q. This size parameter of the nucleon is usually chosen smaller than the known charge radius of the proton, since the experimental radius includes not only the extension b of the quark core of the nucleon, but also the extension of the surrounding pion cloud. Then the masses of the pion cloud surrounding the nucleon and the delta are computed. They depend on the size parameter b. From (10.32), the values for the remaining two parameters α_S and a^{conf} are then determined. The pion–nucleon coupling constant $g^2_{NN\pi}/4\pi$ is chosen to be the one determined from the pion–nucleon interaction. The σ–nucleon coupling constant is fixed to a value of the ^1S or ^3S phase shifts of the nucleon–nucleon scattering at one energy. A reasonable set of parameters for the quadratic and the linear confinement is given in Table 10.2 (see also [10.6,22]).

Table 10.2. Two sets of interaction parameters that satisfy the conditions (10.32). See [10.4,6]

	lin. confinement	quadr. confinement
b (fm)	0.50	0.50
m_q (MeV)	350.0	350.0
α_S	1.12	0.9
a^{lin} (MeV/fm)	133.07	—
a^{quad} (MeV/fm^2)	—	3.35
$g^2_{NN\pi}$	0.0	14.2
g^2_σ	0.0	6.2

10.4 Remarks on the Code

10.4.1 General Structure of the Program

The code is a direct realization of the linear inhomogeneous equation (10.17). The code consist of two parts.

In the first part of the code, all the kernels for the various interactions (confinement, OGE, pion exchange, etc.) are calculated. All interaction diagrams are products of matrix elements in orbital space multiplied with matrix elements in spin–isospin–color (STC) space and multiplied with the corresponding multiplicity. Adding up the two direct diagrams and the five exchange diagrams, we obtain the kernel for a specific interaction. In the code, the order of these seven diagrams is the same as in Table 10.1.

The six-quark matrix elements in STC space have been calculated previously and are stored in the subroutines `SFCME` and `COFP`. The subroutine `SFCME` contains the matrix elements for all interactions, except pion exchange, for the two cases $L = 0$, $S = 0$, and $L = 0, S = 1$ for the three channels NN, $\Delta\Delta$ and *hidden color–hidden color*. These matrix elements are listed also in [10.4]. On the other hand, the subroutine `COFP` contains the matrix elements only for the nucleon-nucleon channel, but for all interactions (including pion exchange) and all four cases $L = \text{even}, \text{odd}$ and $S = 0, 1$. We will need only the matrix elements for the case $L = 0$. The analytic expressions for the six-quark matrix elements in orbital space are listed in [10.23,24]. These expressions are programmed in the present code in several subroutines. In this way, all the following kernels are calculated:

- norm,
- total kinetic energy with center-of-mass kinetic energy removed,
- confinement (either linear or quadratic),
- color–Coulomb kernel,
- color–delta kernel,
- color–magnetic kernel,
- pion exchange at the quark level,
- OBE exchange at the baryon level.

The full Hamiltonian kernel is obtained by summing the kernels of all potentials.

Finally, the two quantities $EINT$ and $RKEY$ are computed. $EINT$ is defined as the Hamiltonian kernel divided by the norm kernel for the outermost mesh point S_N. $RKEY$ is the kinetic energy still contained in the relative motion of the clusters at this last mesh point S_N. This means $EINT - RKEY$ is two times the internal energy of a single nucleon.

In the second part of the program, the RGM equation is solved. The first important step is to compute the complex coefficients p_i and s_i of the ansatz (10.8–11). Since some of the coefficients p_i can become rather large, it is advantageous not to solve the inhomogeneous linear equation (10.17) directly, but the following equivalent equation:

$$\sum_{j=1}^{N-1} \frac{1}{p_i} \tilde{\mathcal{L}}_{ij}{}^{L} \frac{1}{p_j} (p_j\, c_j) = \frac{1}{p_i} \mathcal{M}_i{}^{L}. \tag{10.33}$$

Here one has to solve for the product $p_i c_i$. All the quantities appearing in (10.33) are complex. Actually, it is this equation (10.33), and not (10.17), that is programmed and solved in the second part of the code.

The energy E appearing in the kernels $\tilde{\mathcal{L}}_{ij}^{L}$ and $\mathcal{M}_i{}^{L}$ [see (10.6,16,20)] is the total energy of the six-quark system. This energy is constant. The total energy E is the sum of the given relative kinetic energy E_{rel} in the center-of-mass frame for which the scattering problem is to be solved and the internal energy $EINT - RKEY$:

$$E = E_{\text{rel}} + EINT - RKEY. \tag{10.34}$$

The corresponding energy E_{lab} in the laboratory frame of an incident nucleon scattered off a nucleon at rest would be $E_{\text{lab}} = 2E_{\text{rel}}$.

Finally, from the solution of the RGM equation (10.33), the S-matrix element S_L for the partial wave L is easily evaluated using (10.21):

$$S_L = \sum_{i=1}^{N} \frac{1}{p_i} (p_i c_i) s_i. \tag{10.35}$$

Since all quantities in (10.35) are complex numbers, the S-matrix element S_L is also a complex number. In our calculation, we have one incoming state and one outgoing state, and therefore the conservation of flux requires

$$|S_L| = 1 = |e^{2i\delta_L}|, \tag{10.36}$$

since there is no imaginary potential. The real number δ_L is the phase shift of the partial wave L. The condition (10.36), that is, the unitarity of the calculated S-matrix element, is not automatically guaranteed by our variational method of solving the scattering problem. However, this is not a weak point but an advantage of the variational method, since one can use the deviations of $|S_L|$ from unity to estimate the accuracy of the calculation. In actual calculations, the deviation should be much less than 1%. Any larger deviation is usually either an indication of an error in the code or of an incorrect choice of parameters.

10.4.2 Input Parameters

Throughout the whole code, we use the following convention for distinguishing between integers, double-precision numbers, and double-precision complex variables. Any variable with a first letter between I and N is integer, any variable with a first letter C or Z is double-precision complex, and all the remaining variables are double precision.

Table 10.3. List of input parameters for calculation of kernels

B	b = size parameter of nucleon in fm
QM	m_q = quark mass of up/down quark in MeV
ALFS	α_S = coupling constant of gluon exchange
ALIN	a^{lin} = lin. confinement strength in MeV/fm
AQUAD	a^{quad} = quad. confinement strength in MeV/fm^2
IOBEP	form of meson exchange
PARP	scaling factor for pion exchange
PARS	scaling factor for sigma exchange
PARW	scaling factor for omega exchange
PARR	scaling factor for rho exchange
NOCHNN	number of channels (1,2,3)
SPIN	spin of NN system (0.0 or 1.0)
NS	number of mesh points
SMIN	first mesh point in fm
DS	separation between the mesh points in fm

The input parameters of the code are listed in Tables 10.4 and 10.5. The parameters for the calculation of the kernel part are listed in Table 10.4, while Table 10.5 contains the parameters needed for solving the RGM equation.

The parameters $b, m_{\mathrm{q}}, \alpha_{\mathrm{S}}, a^{\mathrm{lin}}, a^{\mathrm{quad}}$ are adjusted to reproduce the mass splitting between the nucleon and Δ and the stability condition of the nucleon as described in the preceding chapter. To switch between linear confinement and quadratic confinement, one of the parameters a^{lin} or a^{quad} must be different from zero, while the other one is zero.

As already mentioned, we are calculating only the $^1\mathrm{S}_0$ and $^3\mathrm{S}_1$ phases. This means that the orbital angular momentum is generally set to zero ($L = 0$), while the spin is controlled by the parameter $SPIN$.

The number of channels is controlled by the parameter $NOCHNN$. Only the values $1, 2, 3$ are allowed corresponding to a one-channel, two-channel, or three-channel calculation, including the NN, $\Delta\Delta$, and *hidden–color* channel, respectively.

The parameter IOBEP needs some more explanation. The allowed values are IOBEP=0,1,2. This parameter controls the meson-exchange forces.

If IOBEP=0, then there is no meson exchange. One has only the short-range forces due to gluon and quark exchange. There are no medium- and long-range forces.

IOBEP=1 switches on in addition to the gluonic forces, the meson-exchange forces of pion, sigma, rho, and omega exchange on the baryon level. The quark structure of the nucleons is absorbed in form factors that take the spatial extension of the nucleons into account. The strength of these four potentials may be scaled by the input parameters PARP, PARS, PARW, and PARR. For PARP=PARS=PARW=PARR=1, the corresponding strengths of the four meson-exchange potentials are: $g_\pi^2/4\pi = 14.2$, $g_\sigma^2/4\pi = 5.66$, $g_\omega^2/4\pi = 4.5$, and $g_\rho^2/4\pi = 0.5$. We refer to the literature for further details [10.22].

Finally, IOBEP=2 is a calculation where, besides the gluonic part, only pion and sigma exchange is taken into account. Here the sigma-meson is again exchanged at the baryon level as before, but the pion exchange now takes place between the quarks. The strength of the sigma potential may be modified by the parameter PARS. The other three parameters PARP, PARW, and PARR are set to zero in the code.

Table 10.4. List of input parameters for solving the RGM equation

RC	R_c = matching radius in fm
NE	number of relative kinetic energies
$E(1), \ldots, E(NE)$	list of rel. kinetic energies in MeV

The values for the generator coordinates (or mesh points) S_i are controlled by the three parameters NS, $SMIN$ and DS according to the formula: $S_i = SMIN + DS * (i-1)$, $(i = 1, \ldots, NS)$.

The input parameters of the second part are listed in Table 10.4. The matching radius RC is the radius R_c of the ansatz (10.8–11), which distinguishes between the inner and outer regions. It has to be chosen so that the interaction between quarks in different clusters can be safely neglected. Finally E(1),...,E(NE) are the center-of-mass relative kinetic energies for which the NN scattering should be calculated.

10.4.3 Output of the Code

The output of the first part of the code consists first of all of the list of the interaction parameters. Then the matrix elements in STC space for the chosen spin are printed. The main part of the output of the first part of the code are the different kernels (10.23). They are printed if $S_i = S_j$. Finally, the values for $EINT$ and $RKEY$ are printed.

In the second part of the code, we calculate for each given scattering energy $E_{\rm rel}$ in the center-of-mass frame the S-matrix element S_L. The output consists of a list of this energies, the corresponding S-matrix elements S_L, the phase shifts δ_L, and the deviations of the S-matrix element from unitarity in percent, that is, $100 * (|S_L| - 1)$. Experimental values for the phase shifts may be found, for example, in [10.25].

10.4.4 Library Routines

In the code we use routines from the NAG library. Below, in Table 10.5, we list all the routines that are used in the code together with a short description of their purpose and where in the code they are used.

Table 10.5. List of library routines

Routine	Library	Purpose
E01BAF E02BBF	NAG NAG	Cubic spline interpolation and evaluation of $V_{\rm OBE}$ between given sets of radii. Used to compute OBEP kernels in numerical integration.
S15AEF	NAG	Computes error function $erf(x)$. Used for linear confinement kernel.
F04ADF	NAG	Solves set of linear equations $Ax = b$ where A = complex matrix, b = complex vector. Used to solve the RGM equation (10.33).

10.4.5 Performance

The code was programmed on a COMPAREX 7/88 computer. The operating system was MVS. The execution time for this version of the code with only central forces is roughly 7 seconds for the kernel part (NS=10) and 2 seconds for the RGM part (NE=5).

10.4.6 Sample Parameters and Exercises

In Table 10.6, we give a set of sample parameters for the calculation of nucleon–nucleon scattering using only the quark–gluon exchange, but no meson exchange. We also list the S-matrix element S_L of the state $^1S_0$ for the kinetic energy $E_{\text{rel}} = 50$ MeV, which is calculated using the listed sample parameters. When running the code, one should first try to reproduce this result.

Table 10.6. Sample parameters for the kernel and RGM part of the code and result for S-matrix element S_L in the state $^1S_0$ calculated with these parameters.

B	0.50
QM	350.0
ALFS	1.12
ALIN	133.07
AQUAD	0.0
IOBEP	0
PARP	0.0
PARS	0.0
PARW	0.0
PARR	0.0
NOCHNN	1
SPIN	0.0
NS	10
SMIN	0.3
DS	0.35
RC	3.0
NE	1
E	50.0
S_L	(0.59443, −0.80394)

The next important check should be to show the independence of the results from the choice of the mesh points S_i and from the choice of the matching point R_c, as long as reasonable values are used. Of course, all the Hamiltonian and norm kernels depend on the mesh points S_i, but the final result, that is, the S-matrix, should be independent of the parameters S_i and R_c.

The mesh points should be chosen dense enough that the expansion of the relative wave functions into the basis functions $\tilde{u}^L$ (10.10) is a good approximation. On the other hand, the set of mesh points should be not too dense, since then the basis functions become linearly dependent and numerical problems occur. Also, the matching radius should be chosen large enough that the interaction between the clusters can be neglected. But again, this radius must not be too large, since this causes numerical problems because the complex parameter p_i gets very large in this case. In practice, there is a wide range of possible values for the mesh points S_i and the matching radius R_c, with which the S-matrix elements are close to unitarity and independent of the choice of these parameters.

Finally, if one understands how these parameters have to be chosen, one can study the influence of the various potentials on the phase shifts. Reasonable questions are, for example:

- Which part of the interaction causes the short-range repulsion, medium-range attraction, long-range attraction?
- Which part of the interaction is responsible for the difference of the nucleon–nucleon phase shifts between the channels $^1S_0$ and $^3S_1$?

One is now able to reproduce the results obtained in [10.3–6,22]. It should be noted, however, that the meson-exchange potentials we are using in the present code are slightly different from the old ones employed in [10.3–6,22]. Therefore, in some cases, small modifications, especially of the strength of the sigma-meson exchange, may be necessary to obtain perfect agreement with those references.

10.5 Technical Note

The code QMODL2 mainly follows the FORTRAN 77 standard, with the exception that it uses complex*16 arithmetic. Due to the required array dimensions, the code is not intended to be run on a microcomputer. A detailed description of the input and output can be found in Sects. 10.4.2 and 10.4.3. For convenience, an input file QMODL2.INP, corresponding to the example values of Table 10.6, is included on the disk. As already stated in Sect. 10.4.4, the code calls some NAG subroutines, which will be briefly described in the following.

S15AEF

Purpose:		To calculate the value of the error function erf(x) via the routine name
Usage:		`REAL FUNCTION S15AEF(X,IFAIL)`
Arguments:		
X	–	Specifies the argument x (input)
IFAIL	–	integer for error detection

E01BAF

Purpose: Cubic-spline interpolation to a given data set

Usage: `CALL E01BAF (M,X,Y,K,C,LCK,WK,LWK,IFAIL)`

Arguments:

- M – Specifies the number of data points (≥4) (input)
- X – Real array of dimension at least M (input)
  Before entry, X(I) must be set to the Ith data value of the independent variable x, for I=1,2,...,M.
- Y – Real array of dimension at least M (input)
  Before entry, Y(I) must be set to the Ith data value of the dependent variable y, for I=1,2,...,M.
- K – Real array of dimension at least LCK (output)
  On exit, K(I) contains the value of the Ith knot provided by `E01BAF`, for I=1,2,...,M+4.
- C – Real array of dimension at least LCK (output)
  On exit, C(I) contains the coefficient of the B-spline $N_I(x)$ (I=1,2,...,M) in the B-spline representation of the interpolating spline.
- LCK – Specifies the smaller of the dimensions of K and C
- WK – Real working space of dimension LWK
- LKW – Dimension of the working space WK (LWK≥6M+16)

E02BBF

Purpose: To evaluate a cubic spline from its B-spline representation

Usage: `CALL E02BBF (NCAP7,K,C,X,S,IFAIL)`

Arguments:

- NCAP7 – Integer that specifies NCAP+7 where NCAP is the number of intervals (one greater than the number of interior knots) over which the spline is defined (NCAP7≥8)
- K – Real array of dimension at least NCAP7 (input)
  Before entry, K(I) must be set to the Ith member of the complete set of knots, for I=1,2,...,NCAP7
- C – Real array of dimension at least NCAP7 (input)
  Before entry, C(I) must be set to the value of the Ith B-spline coefficient, for I=1,2,...,NCAP+3
- X – Real number that specifies the argument at which the cubic spline is to be evaluated (input)
- S – Contains the value of the spline (output)
- IFAIL – integer for error detection

F04ADF

Purpose:		To calculate the approximate solution of a set of complex linear equations with multiple right-hand sides by Crout's factorization method
Usage:		`CALL F04ADF (A,IA,B,IB,N,M,C,IC,WK,IFAIL)`
Arguments:		
A	–	Complex array of dimension (IA,R), R≥N (input) Before entry, A must contain the complex matrix On exit, A contains the Crout factorization
IA	–	First dimension of array A, IA≥N (input)
B	–	Complex array of dimension (IB,Q), where Q≥N Before entry, B must contain the elements of the M right-hand sides stored in columns
IB	–	First dimension of array B, IB≥N (input)
N	–	Order of matrix A (input)
M	–	Specifies the number of right-hand sides (input)
C	–	Complex array of dimension (IC,R), where R≥M On exit, C contains the M complex solution vectors
IC	–	First dimension of array C, IC≥N (input)
WK	–	Real array of dimension at least N, used as working space
IFAIL	–	integer for error detection

References

[10.1] N. Isgur, G. Karl, Phys. Rev. D**19**, 2653 (1979); N. Isgur, G. Karl, R. Koniuk, Phys. Rev. D**25**, 2394 (1982)

[10.2] G. Karl, Progr. Part. Nucl. Phys. **13**, 243 (1985)

[10.3] A. Faessler, F. Fernandez, G. Lübeck, K. Shimizu, Phys. Lett. **112B**, 201 (1982)

[10.4] A. Faessler, F. Fernandez, G. Lübeck, K. Shimizu, Nucl. Phys. A**402**, 555 (1983)

[10.5] A. Faessler, F. Fernandez, Phys. Lett. **124B**, 145 (1983)

[10.6] K. Bräuer, A. Faessler, F. Fernandez, K. Shimizu, Z. Phys. A**320**, 609 (1985)

[10.7] A. Faessler, U. Straub, Phys. Lett. **183B**, 10 (1987)

[10.8] U. Straub, Z. Zhang, K. Bräuer, A. Faessler, S. B. Khadkikar, G. Lübeck, Nucl. Phys. A**483**, 686 (1988)

[10.9] M. Oka, K. Yazaki, Progr. Theor. Phys. **66**, 551 and 572 (1981)

[10.10] M. Oka, K. Shimizu, K. Yazaki, Phys. Lett. **130B**, 365 (1983)

[10.11] U. Straub, Z. Zhang, K. Bräuer, A. Faessler, S. B. Khadkikar, Phys. Lett. **B200**, 241 (1988)

[10.12] K. Erkelenz, Phys. Rep. **13**, 191 (1974)

[10.13] K. Holinde, Phys. Rep. **68**, 121 (1981)

[10.14] K. Holinde, R. Machleidt, Nucl. Phys. A**256**, 479 (1978)

[10.15] Z. Zhang, K. Bräuer, A. Faessler, K. Shimizu, Nucl. Phys. A**443**, 557 (1985)
[10.16] C.S. Warke, R. Shanker, Phys. Rev. C**21**, 2643 (1980)
[10.17] K. Kamimura, Prog. Theor. Phys. Suppl. **62**, 236 (1977)
[10.18] W. Kohn, Phys. Rev. **74**, 1763 (1948)
[10.19] L. Hulthén, Ark. Mat. Astr. Fys. **35A**, No. 25 (1948)
[10.20] T. Kato, Prog. Theor. Phys. **6**, 394 (1951)
[10.21] A. de Rújula, H. Georgi, S.L. Glashow, Phys. Rev. D**12**, 147 (1975)
[10.22] A. Faessler, Prog. Part. Nucl. Phys. **13** 253–273 (1985)
[10.23] G. Lübeck, Diplomarbeit, Universität Tübingen (1982)
[10.24] K. Bräuer, Ph.D. Thesis, Universität Tübingen (1985)
[10.25] R.A. Arndt, L.D. Roper, R.A. Bryan, R.B. Clark, B.J. Ver West, K. Signell, Phys. Rev. D**28**, 97 (1983)

11. Hadron–Hadron and Hadron–Nucleus Scattering

F. Lenz and D. Stoll

11.1 Introduction

In this contribution we discuss two problems in the physics of strong interactions. These problems are intended to introduce the reader to two important and active areas of research in the field of theoretical nuclear physics.

In the first part, the multiple-scattering formalism of nuclear reactions [11.1,2] will be introduced and actual calculational methods will be presented. The multiple-scattering formalism constitutes one of the basic methods employed in the traditional areas of medium- and high-energy nuclear physics. It has been applied to the description of exotic atoms ($\pi^-, K^-, \bar{p}$-atoms) [11.2,3], the resonance-dominated π-nucleus scattering [11.4], and proton–nucleus scattering for incident proton energies ranging from a few hundred MeV to tens of GeV [11.5–7]; a whole variety of reactions has been treated, such as elastic scattering, excitation of discrete nuclear states, one- and many-nucleon knockout reactions, and coherent and incoherent production processes. We will illustrate the classical computational method, based on the optical-potential approximation and demonstrate the applicability of stochastic methods, which might allow for new approaches to multiple-scattering calculations. This discussion will be carried out in the context of the conceptually simplest case of multiple scattering, the scattering on a system of fixed scatterers. By the assumption of infinitely heavy nucleons, projectile–nucleus scattering is, for a fixed configuration of nucleons, reduced to a many-center-potential scattering problem. Nevertheless, the many-body aspect has not been lost entirely. The final projectile–nucleus amplitude is obtained only after averaging over all configurations with the weight given by the nuclear ground-state density. It is this last step that in the more traditional approaches to multiple scattering is performed only approximatively. In the optical-potential approximation, for example, instead of averaging the scattering amplitudes corresponding to different configurations one calculates the scattering on an appropriately defined averaged configuration. By employing stochastic methods, this configurational averaging can be performed without such approximations [11.8]. We will provide a program that offers the user a computational study of these two methods for the treatment of multiple-scattering processes for the case of projectile–^{4}He scattering.

The second part of this contribution addresses the problem of the description of hadronic interactions in terms of quark degrees of freedom. While the theory of strong interactions, quantumchromodynamics (QCD), is the theory of interacting quarks and gluons, nuclear structure and reactions are traditionally described by hadrons, that is, nucleons, Δ-resonances, pions, and so on. One of the fundamental aims of nuclear theory is to understand the relation between the phenomenological hadronic description and the basic quark–gluon theory. In the context of single-hadron spectroscopy, the quark model has been crucial in clarifying this relation [11.9] – QCD was actually developed on the basis of the success of the quark model. It is hoped that the application of similar ideas may provide a deeper understanding of multi-hadron systems as well. There have been numerous attempts at a detailed numerical explanation of hadronic interactions, in particular, of the NN interaction, in the framework of non-relativistic quark models [11.10,11]. We do not review these efforts, nor do we illustrate the main technical tool used in these calculations, the so-called "resonating group method," which was developed in the context of low-energy reactions (however, see Chapts. 4 and 10). Rather the emphasis here is on illustrating the new class of problems encountered and novel models developed in the attempt to formulate a theory of interacting composite systems, when the constituents are confined. Investigations of such theories [11.12] have led to the discovery of a class of few-body problems, which, by employing techniques reminiscent of those used in electrostatics, are amenable to exact numerical solutions. We will focus in our discussion on these novel analytical and numerical methods. The physical interest of such investigations of exactly solvable few-body problems is the emergence of the complex behavior of composite hadronic systems on the basis of simple dynamics of the confined constituents. The program included allows the user to study the exact many-channel S-matrix corresponding to this four-body problem.

11.2 Multiple Scattering Off Nuclei

11.2.1 General Formalism

We start our discussion with an introduction to the general multiple-scattering formalism [11.1,2] and give a description of the main lines of research. Underlying this formalism is the impulse approximation, in which scattering of a projectile on a bound nucleon is approximately identified with the scattering on a single free nucleon. Projectile–nucleus scattering in turn is described as a sequence of such single quasi-free scattering events.

Our formal development starts with the Hamiltonian H for the projectile–nucleus system:

$$H = H_0 + \sum_{i=1}^{N} v_i; \qquad H_0 = H_A + K\,. \tag{11.1}$$

H_0 contains the nuclear Hamiltonian H_A and the projectile kinetic-energy operator K; v_i describes the interaction of the projectile with nucleon i. The projectile–nucleus transition operator T satisfies the Lippmann–Schwinger equation:

$$T = \sum_i v_i + \sum_i v_i G_0 T \,. \tag{11.2}$$

G_0 is the Green's function of the projectile–nucleus system in the absence of projectile–nucleon interactions:

$$G_0 = (E - H_0 + \mathrm{i}\eta)^{-1} \,. \tag{11.3}$$

In the definition of G_0, the "outgoing-wave" boundary condition is implemented ($\eta > 0$). The transition operator T acts on both projectile and nuclear degrees of freedom; specific transition amplitudes are obtained by projection on definite projectile plane-wave states and nuclear initial and final states. In particular, elastic scattering is given by the matrix element:

$$T_0(\boldsymbol{k}', \boldsymbol{k}) := \langle 0, \boldsymbol{k}' | T | \boldsymbol{k}, 0 \rangle \tag{11.4}$$

with incident and projectile outgoing momenta $\boldsymbol{k}$ and $\boldsymbol{k}'$, respectively, and with the nuclear ground state $|0\rangle$. The projectile–nucleon interaction v_i is now eliminated in favor of a quantity that is more closely related to the free projectile–nucleon t-matrix. The "bound" projectile–nucleon t-matrix is defined via the integral equation

$$\tau_i = v_i + v_i G_0 \tau_i , \tag{11.5}$$

with the formal solution

$$\tau_i = (1 - v_i G_0)^{-1} v_i \,. \tag{11.6}$$

Indeed, τ_i reduces to the free projectile–nucleon t-matrix if H_0 can be replaced by the sum of kinetic-energy operators of nucleon i and the projectile. Owing to the presence of the many-body Green's function G_0, τ_i is a complicated many-body operator, which contains the influence of the binding forces on the projectile–nucleon scattering.

The Lippmann–Schwinger equation (11.2) is rewritten through the following formal definitions and identities. First, T is written as a sum over subamplitudes T_i defined by

$$T = \sum_i^A T_i \,; \qquad T_i = v_i + v_i G_0 \sum_j^A T_j \,. \tag{11.7}$$

T_i describes a scattering event in which the last scattering has occurred with nucleon i. Rewriting (11.7) with the help of the bound projectile–nucleon amplitudes (11.6), we have

$$(1 - v_i G_0) T_i = v_i + v_i G_0 \sum_{j \neq i} T_j = (1 - v_i G_0) \left(\tau_i + \tau_i G_0 \sum_{j \neq i} T_j \right) , \tag{11.8}$$

and therefore

$$T_i = \tau_i + \tau_i G_0 \sum_{j \neq i} T_j \,. \tag{11.9}$$

The interaction with projectile i has been summed into the bound transition operator τ_i. The summation restriction $j \neq i$ occurring in (11.9) prevents any further immediate interaction with nucleon i. Equations (11.5,7,9) represent the multiple-scattering formulation of the Lippmann–Schwinger equation (11.2). By formal iteration of (11.9), the multiple-scattering expansion of the projectile–nucleus transition operator is obtained:

$$T = \sum_i \tau_i + \sum_i \tau_i G_0 \sum_{j \neq i} \tau_j + \sum_i \tau_i G_0 \sum_{j \neq i} \tau_j G_0 \sum_{k \neq j} \tau_k + \dots \,. \tag{11.10}$$

With the multiple-scattering formulation we have achieved a separation of the effects of the nuclear dynamics – the binding corrections occurring in the definition of τ_i (11.5) – from the dynamics associated with the repeated interaction of the projectile with the different nucleons (11.9). This separation is crucial for developing approximation schemes that properly account for the specific dynamics of the particular projectile–nuclear systems. Most of the applications of the multiple-scattering formalism are based on the fixed-scatterer approximation. In this approximation, the complicated many-body transition operator τ_i is replaced by an operator which, apart from implementing momentum conservation, acts in a non-trivial way only on the projectile states. In terms of its momentum-space matrix elements, this approximate transition operator is expressed as

$$\langle \boldsymbol{k}', \boldsymbol{p}_i' | \tau_i | \boldsymbol{k}, \boldsymbol{p}_i \rangle \approx \tau_i(E, \boldsymbol{k}' - \boldsymbol{k}) \delta\left[\boldsymbol{k}' + \boldsymbol{p}_i' - (\boldsymbol{k} + \boldsymbol{p}_i)\right] , \tag{11.11}$$

with $\boldsymbol{p}_i, \boldsymbol{p}_i'$ the initial and final ith nucleon momenta. The function $\tau_i(E, \boldsymbol{k}' - \boldsymbol{k})$ is in general identified with the free projectile–nucleon amplitude. We have assumed here for simplicity that this amplitude depends on the momentum transfer only.

The fixed-scatterer approximation (11.11) is valid only if the nuclear Hamiltonian in the defining equation of the τ-operator (11.5) can be replaced by the kinetic-energy operator of the ith nucleon, that is, if binding effects are negligible. This requires the projectile–nucleon interaction time to be short in comparison with typical nuclear times. It is very hard to assess quantitatively the validity of this approximation; judging from its success, the impulse approximation seems to be justified for high-energy proton–nucleus interactions, while in resonance-dominated pion–nucleus scattering the fixed-scatterer approximation fails. Single-particle energies of the order of 10–20 MeV are not negligible on the scale of the half-width of 50 MeV of the 3–3 resonance. This important difference between pion scattering off a free and a bound nucleon was the starting point in the development of the "isobar-doorway" formalism [11.4] which provided the framework for a quantitative understanding of the basic pion–nucleus reactions.

Here we do not follow further this line of investigation but rather concentrate on the multiple-scattering formulation within the fixed-scatterer approximation (11.11). In the majority of applications of the multiple-scattering formalism, as a further approximation, the many-body Green's function in the multiple-scattering equations (11.9) is replaced by the projectile (one-body) Green's function

$$G_0 \approx (E - K + \mathrm{i}\eta)^{-1} \ . \tag{11.12}$$

The fixed-scatterer approximation (11.11) does not imply approximation (11.12); the two approximations concern, on the one hand, the dynamics of the single-scattering event (11.11) and the dynamics of the multiple-scattering (11.9,12) on the other. The effect of the combined approximations (11.11,12) is that the subamplitudes T_i become diagonal in the nucleon coordinates. This allows us to define operators that act on the projectile degrees of freedom only and depend parametrically on the nucleon coordinates. We first calculate the configuration-space matrix elements by Fourier transforming (11.11):

$$\langle \boldsymbol{r}', \boldsymbol{r}_i' | \tau_i | \boldsymbol{r}, \boldsymbol{r}_i \rangle = \tilde{\tau}_i(\boldsymbol{r} - \boldsymbol{r}_i)\delta(\boldsymbol{r}' - \boldsymbol{r})\,\delta(\boldsymbol{r}_i - \boldsymbol{r}_i') \, , \tag{11.13}$$

where $\tilde{\tau}$ is the Fourier transform of the amplitude

$$\tilde{\tau}_i(\boldsymbol{r} - \boldsymbol{r}_i) = \int \mathrm{d}^3 q \exp\left[\mathrm{i}\boldsymbol{q}(\boldsymbol{r} - \boldsymbol{r}_i)\right] \tau_i(E, \boldsymbol{q}), \tag{11.14}$$

and the local operator $\hat{\tau}$ is defined by

$$\delta(\boldsymbol{r} - \boldsymbol{r}')\tilde{\tau}_i(\boldsymbol{r} - \boldsymbol{r}_i) = \langle \boldsymbol{r}' | \hat{\tau}_i(\boldsymbol{r}_i) | \boldsymbol{r} \rangle \, . \tag{11.15}$$

Projecting the multiple-scattering equation (11.9) onto nuclear states with the nucleons fixed at the positions $\boldsymbol{r}_i$, one obtains with

$$\begin{aligned} &\langle \boldsymbol{r}_1', \boldsymbol{r}_2', \dots, \boldsymbol{r}_A' | T_i | \boldsymbol{r}_1, \boldsymbol{r}_2, \dots, \boldsymbol{r}_A \rangle \\ &= \hat{T}_i(\boldsymbol{r}_1, \boldsymbol{r}_2, \dots, \boldsymbol{r}_A) \prod_{l=1}^{A} \delta(\boldsymbol{r}_l - \boldsymbol{r}_l') \end{aligned} \tag{11.16}$$

the following operator equation in the Hilbert space of projectile states (omitting the parametric dependence on the nucleon coordinates):

$$\hat{T}_i = \hat{\tau}_i + \hat{\tau}_i \frac{1}{E - K + \mathrm{i}\eta} \sum_{j \neq i}^{A} \hat{T}_j . \tag{11.17}$$

The projectile–nucleus elastic amplitude is given by [cf. (11.4,7)]

$$\begin{aligned} T_0(\boldsymbol{k}, \boldsymbol{k}') \;=\; & \sum_{i=1}^{A} \int \mathrm{d}^3 r_1, \dots, \mathrm{d}^3 r_A \rho(\boldsymbol{r}_1, \dots, \boldsymbol{r}_A) \\ & \times \langle \boldsymbol{k}' | \hat{T}_i(\boldsymbol{r}_1, \dots, \boldsymbol{r}_A) | \boldsymbol{k} \rangle , \end{aligned} \tag{11.18}$$

with the nuclear ground-state density ρ and wave function ϕ:

$$\begin{aligned} \rho(\boldsymbol{r}_1,\ldots,\boldsymbol{r}_A) &= |\phi_0(\boldsymbol{r}_1,\ldots,\boldsymbol{r}_A)|^2; \\ \phi_0(\boldsymbol{r}_1,\ldots,\boldsymbol{r}_A) &= \langle \boldsymbol{r}_1,\ldots,\boldsymbol{r}_A|0\rangle . \end{aligned} \tag{11.19}$$

Thus, in the fixed-scatterer approximation, the projectile–nucleus scattering amplitude is obtained by solving the multiple-scattering equation (11.17) for an arbitrary configuration of nucleons and by configurational averaging of these amplitudes with the weight given by the ground-state density (11.18).

Despite the rather severe approximations already necessary for deriving (11.17), the amplitude (11.18) is in general not evaluated exactly. Rather, these equations serve as starting points for further approximations. In the one most commonly used, the so-called optical-potential approximation, instead of averaging the amplitudes for the individual configurations, one calculates the amplitude for an averaged configuration. Projecting (11.17) onto the nuclear ground-state wave function and approximating

$$\begin{aligned} &\Sigma_{i=1}^{A} \int \mathrm{d}^3r_1,\ldots,\mathrm{d}^3r_A \rho(\boldsymbol{r}_1,\ldots,\boldsymbol{r}_A)\hat{\tau}_i(\boldsymbol{r}_i)\frac{1}{E-K+\mathrm{i}\eta}\sum_{j\neq i}\hat{T}_j(\boldsymbol{r}_1,\ldots,\boldsymbol{r}_A) \\ &\approx \quad U(\boldsymbol{r})\frac{1}{E-K+\mathrm{i}\eta}\overline{T}, \end{aligned} \tag{11.20}$$

with

$$\overline{T} = \sum_i \int \mathrm{d}^3r_1,\ldots,\mathrm{d}^3r_A \hat{T}_i(\boldsymbol{r}_1,\ldots,\boldsymbol{r}_A)\rho(\boldsymbol{r}_1,\ldots,\boldsymbol{r}_A) \tag{11.21}$$

and

$$U(\boldsymbol{r}) = \frac{A-1}{A}\sum_i \int \mathrm{d}^3r_1,\ldots,\mathrm{d}^3r_A \rho(\boldsymbol{r}_1,\ldots,\boldsymbol{r}_A)\hat{\tau}(\boldsymbol{r}_i), \tag{11.22}$$

we see that the transition operator $\overline{T}$ for elastic scattering satisfies a standard one-body Lippmann–Schwinger equation

$$\overline{T} = \frac{A}{A-1}U + U\frac{1}{E-K+\mathrm{i}\eta}\overline{T}, \tag{11.23}$$

with the optical-potential operator U. In configuration space [cf. (11.15)] U is local and obtained by folding the ground-state density with the projectile–nucleon amplitude:

$$\begin{aligned} \langle \boldsymbol{r}'|U|\boldsymbol{r}\rangle &= \frac{A-1}{A}\delta(\boldsymbol{r}-\boldsymbol{r}') \\ &\times \sum_{i=1}^{A} \int \mathrm{d}^3r_1,\ldots,\mathrm{d}^3r_A \rho(\boldsymbol{r}_1,\ldots,\boldsymbol{r}_A)\tilde{\tau}_i(\boldsymbol{r}-\boldsymbol{r}_i). \end{aligned} \tag{11.24}$$

In the approximation (11.20), only the nuclear ground state appears as an intermediate state between two scatterings and therefore the optical potential (11.22) is the ground-state average of the potentials corresponding to individual configurations.

At high energies, no restriction to intermediate nuclear states is necessary. In Glauber theory, the essential simplification arises by replacing the projectile propagator in (11.17) by the eikonal propagator

$$\begin{aligned}\left\langle r' \left| \frac{1}{E-K+\mathrm{i}\eta} \right| r \right\rangle &= -\frac{2m}{4\pi\hbar^2}\frac{\mathrm{e}^{\mathrm{i}k|\mathbf{r}-\mathbf{r}'|}}{|\boldsymbol{r}-\boldsymbol{r}'|} \\ &\approx -\mathrm{i}\frac{m}{\hbar^2 k}\mathrm{e}^{\mathrm{i}k(z'-z)}\theta(z'-z)\delta^2(\boldsymbol{b}-\boldsymbol{b}'), \end{aligned} \tag{11.25}$$

which allows only for forward straight-line propagation along the z-axis, the direction of the incident projectile momentum $\boldsymbol{k}$. $\boldsymbol{b}$ is the projection of $\boldsymbol{r}$ onto the impact-parameter plane perpendicular to z [$\boldsymbol{r} = (z, \boldsymbol{b})$]. Furthermore, at high energies, projectile–nucleon scattering occurs predominantly at small angles, where the momentum transfer is orthogonal to the incident direction, and thus the elementary amplitude (11.14) can be approximated by

$$\begin{aligned}\tilde{\tau}_i(\boldsymbol{r}-\boldsymbol{r}_i) &\approx \int_{-\infty}^{\infty} \mathrm{d}q_z \mathrm{e}^{-\mathrm{i}q_z(z-z_i)} \int \mathrm{d}^2 q_\perp \mathrm{e}^{-\mathrm{i}\mathbf{q}_\perp(\mathbf{b}-\mathbf{b}_i)}\tau_i(E, q_\perp) \\ &= \delta(z-z_i)\gamma_i(\boldsymbol{b}-\boldsymbol{b}_i). \end{aligned} \tag{11.26}$$

As a consequence, the impact parameter $\boldsymbol{b}$ is not changed in the course of the multiple scattering through the nucleus; defining correspondingly

$$\langle \boldsymbol{r}'|\hat{T}_i|\boldsymbol{r}\rangle = \delta(\boldsymbol{b}-\boldsymbol{b}')T_i(z', z; \boldsymbol{b}), \tag{11.27}$$

we find that the multiple-scattering equation (11.17) reads in coordinate space

$$\begin{aligned}T_i(z', z; \boldsymbol{b}) &= \delta(z-z_i)\delta(z'-z_i)\gamma_i(\boldsymbol{b}-\boldsymbol{b}_i) + \frac{m}{\mathrm{i}\hbar^2 k}\int_{-\infty}^{\infty} \mathrm{d}z''\delta(z'-z_i) \\ &\quad \times\gamma_i(\boldsymbol{b}-\boldsymbol{b}_i)\mathrm{e}^{-\mathrm{i}k(z'-z'')}\theta(z'-z'')\sum_{j\neq i} T_j(z'', z; \boldsymbol{b}), \end{aligned} \tag{11.28}$$

which becomes an algebraic equation for the quantities $\Gamma_i(\boldsymbol{b})$ defined by

$$\mathrm{e}^{\mathrm{i}kz'}T_i(z', z; \boldsymbol{b})\mathrm{e}^{-\mathrm{i}kz} = \delta(z'-z_i)\Gamma_i(z, \boldsymbol{b}), \tag{11.29}$$

$$\begin{aligned}\Gamma_i(\boldsymbol{b}) = \int \mathrm{d}z\Gamma_i(z, \boldsymbol{b}) &= \gamma_i(\boldsymbol{b}-\boldsymbol{b}_i) \\ &+ \frac{m}{\mathrm{i}\hbar^2 k}\gamma_i(\boldsymbol{b}-\boldsymbol{b}_i)\sum_{j\neq i}\theta(z_i-z_j)\Gamma_j(\boldsymbol{b}). \end{aligned} \tag{11.30}$$

In terms of Γ_i, the nuclear amplitude is given by

$$T_0 \approx \int \mathrm{d}^2 b \mathrm{e}^{\mathrm{i}\mathbf{b}(\mathbf{k}-\mathbf{k}')} \int \mathrm{d}^3 r_1, \ldots, \mathrm{d}^3 r_A \rho(\boldsymbol{r}_1, \ldots, \boldsymbol{r}_A)\sum_i \Gamma_i(\boldsymbol{b}; \boldsymbol{r}_1, \ldots, \boldsymbol{r}_A). \tag{11.31}$$

The combination of approximations (11.25) and (11.26) leads to a multiple-scattering expansion, the Born series of (11.30) with a finite number of terms; the summation restriction in (11.17) together with the restriction to forward propagation does not allow the projectile to be scattered back to a nucleon hit earlier. The advantage of the Glauber approximation is the simplicity in the evaluation of the scattering amplitude, which for light nuclei and appropriate choices for ground-state densities and projectile–nucleon amplitude is possible analytically. The application has been very successful for high-energy proton–nucleus scattering. Furthermore, extension to inelastic processes is straightforward. In particular, Glauber theory provides the only practical framework for the systematic description of inclusive processes. On the other hand, the underlying assumptions clearly restrict reasonable application of Glauber theory to reactions at very high energies, where the inherent small-angle approximation is valid.

Outside the regime of validity of the Glauber approximation, applicability and further improvements in the multiple-scattering formalism are limited for practical reasons to processes where one or at most two nucleons are actively involved. In pion–nucleus scattering, for example, the most complex reactions that have been described are, on the one hand, quasi-free scattering with the final state consisting of a pion, a knocked-out nucleon, and the residual nucleus [11.13], and, on the other hand, double-charge-exchange reactions to bound nuclear states [11.14]. Similarly, the development of the optical potential has been restricted to such basic two-nucleon processes. These technical limitations in the theoretical description contrast with the well-known fact that, at medium and high energies, these simple reactions constitute only a small fraction of the total reaction cross section. In most of the reactions, several nucleons are knocked out. Our theoretical tools are not adequate for describing even the corresponding summed cross sections, nor is the feedback of such multinucleon processes on simple reactions like elastic scattering understood. Progress in the theoretical treatment of medium- and high-energy reactions, and in particular the extension of present-day calculations to multinucleon processes, requires above all the development of new numerical approaches. We now describe such an approach and demonstrate the practical possibility for obtaining exact numerical solutions to the multiple scattering equation (11.17). We rewrite this equation by defining states $|\psi_k^i\rangle$

$$\hat{\tau}_i|\psi_k^i\rangle = \hat{T}_i|\boldsymbol{k}\rangle\,, \tag{11.32}$$

which according to (11.17) satisfy the equation

$$|\psi_k^i\rangle = |\boldsymbol{k}\rangle + \frac{1}{E-K+\mathrm{i}\eta}\sum_{j\neq i}\hat{\tau}_j|\psi_k^j\rangle\,. \tag{11.33}$$

For the following calculation, we assume a zero-range projectile–nucleon t-matrix (11.15),

$$\tilde{\tau}_i(\boldsymbol{r}-\boldsymbol{r}_i) = \frac{-4\pi\hbar^2}{2m}a\delta(\boldsymbol{r}-\boldsymbol{r}_i), \tag{11.34}$$

where $a = a(E)$ is the (in general complex) projectile–nucleon isotropic scattering amplitude (in the laboratory system). The scattering amplitude for a given configuration of fixed nucleons is given by

$$f_{\mathbf{k}',\mathbf{k}}(\mathbf{r}_1,\ldots,\mathbf{r}_A) = \frac{-2m}{4\pi\hbar^2}\left\langle \mathbf{k}'\left|\sum_i \hat{T}_i\right|\mathbf{k}\right\rangle = \sum_i a\mathrm{e}^{-\mathrm{i}\mathbf{k}'\mathbf{r}_i}\psi^i_{\mathbf{k}}(\mathbf{r}_i)\,, \tag{11.35}$$

with

$$\psi^i_{\mathbf{k}}(\mathbf{r}_i) = \psi^i_{\mathbf{k}}(\mathbf{r})\Big|_{\mathbf{r}=\mathbf{r}_i} = \left\langle \mathbf{r}|\psi^i_{\mathbf{k}}\right\rangle\Big|_{\mathbf{r}=\mathbf{r}_i}\,. \tag{11.36}$$

Projecting (11.33) onto configuration space and identifying the projectile coordinate as in (11.36) with the appropriate nucleon coordinate yields the following algebraic system of equations:

$$\psi^i_{\mathbf{k}}(\mathbf{r}_i) = \mathrm{e}^{\mathrm{i}\mathbf{k}\mathbf{r}_i} + a\sum_{j\neq i}^{N}\frac{\mathrm{e}^{\mathrm{i}k|\mathbf{r}_j-\mathbf{r}_i|}}{|\mathbf{r}_j-\mathbf{r}_i|}\psi^j_{\mathbf{k}}(\mathbf{r}_j)\,. \tag{11.37}$$

Thus, scattering on a given configuration of nucleons requires solution of the algebraic system (11.37). In comparison with the Glauber result (11.30), here the multiple-scattering series does not terminate. By solving the system of equations (11.37), multiple-scattering processes to infinite order are included. The substitutions

$$\gamma_i = \frac{-2m}{4\pi\hbar^2 a}\delta(\mathbf{b}-\mathbf{b}_i); \qquad \Gamma_i(\mathbf{b}) = \frac{-2m}{4\pi\hbar^2 a}\delta(\mathbf{b}-\mathbf{b}_i)\mathrm{e}^{-\mathrm{i}\mathbf{k}\mathbf{r}_i}\psi^i_{\mathbf{k}}(\mathbf{r}_i)\,,$$

together with the eikonal approximation to the propagator, transform (11.37) into (11.30). Scattering off the nucleus is obtained by averaging these individual amplitudes with the weight given by the ground-state density [cf. (11.18,35)]

$$f(\mathbf{k}',\mathbf{k}) = \int \mathrm{d}^3r_1,\ldots,\mathrm{d}^3r_A\rho(\mathbf{r}_1,\ldots,\mathbf{r}_A)f_{\mathbf{k}',\mathbf{k}}(\mathbf{r}_1,\ldots,\mathbf{r}_A). \tag{11.38}$$

The results of a stochastic evaluation [cf. (11.38)] will be compared with an optical-model calculation in which the scattering amplitude is obtained by solving the radial Schrödinger equation

$$\left[-\frac{\mathrm{d}^2}{\mathrm{d}r^2} + \frac{l(l+1)}{r^2} + u(r)\right]y_l(r) = k^2 y_l(r), \tag{11.39}$$

where according to (11.24) and the zero-range parametrization (11.34) the optical potential $u(r)$ is given by

$$u(r) = 4\pi a(A-1)\rho(r) \tag{11.40}$$

with the one-body ground-state density ρ. The radial wave functions $y_l(r)$ are regular at the origin ($\propto r^{l+1}$), and they determine the partial-wave amplitudes f_l via the Wronskian W [11.15]:

$$f_l = \frac{-1}{k} \frac{W\left(rj_l, y_l\right)}{W\left(rh_l^{(+)}, y_l\right)}\Bigg|_{r=R} \frac{A}{A-1}, \tag{11.41}$$

where R is chosen so that $u(R)$ (11.40) can be neglected. The final scattering amplitude is

$$f(\boldsymbol{k}', \boldsymbol{k}) = \sum_l^{\infty}(2l+1) f_l P_l(\cos\theta), \tag{11.42}$$

where the Bessel and Hankel functions and the Legendre polynomials are defined according to [11.15].

11.2.2 Numerical Methods

Stochastic Evaluation

The evaluation of the scattering amplitude $f_{\mathbf{k}',\mathbf{k}}$ (11.35) for a given configuration requires solution of the system of N linear equations (11.37), which is performed using the Gauss algorithm [11.16]. The configurational averaging (11.38) is carried out with the help of the Metropolis algorithm, which we use in the form described in [11.17]. For ^{4}He, with center-of-mass fixed at the origin, the ground-state density $\rho(\boldsymbol{x}_1, \boldsymbol{x}_2, \boldsymbol{x}_3)$ is a function of three independent position vectors and is related to the ground-state density defined in (11.19) by

$$\rho(\boldsymbol{x}_1, \ldots, \boldsymbol{x}_4) = \delta(\sum_{i=1}^{4} \boldsymbol{x}_i)\rho(\boldsymbol{x}_1, \boldsymbol{x}_2, \boldsymbol{x}_3). \tag{11.43}$$

For performing the stochastic integration, we construct an auxiliary density $\sigma(\boldsymbol{x}_1, \boldsymbol{x}_2, \boldsymbol{x}_3)$, which on the one hand resembles the exact density ρ and on the other can be integrated analytically. We choose

$$\begin{aligned} \sigma(\boldsymbol{x}_1, \boldsymbol{x}_2, \boldsymbol{x}_3) \prod_{i=1}^{3} \mathrm{d}^3 x_i &= \prod_{\alpha=1}^{9} \sigma_\alpha(y_\alpha) \mathrm{d}y_\alpha \\ &= \prod_{k=1}^{3} \left(\frac{2}{c^2} r_k \mathrm{e}^{-r_k^2/c^2} \mathrm{d}r_k\right) \left(\frac{\sin\theta_k \mathrm{d}\theta_k}{2}\right) \frac{\mathrm{d}\phi_k}{2\pi}, \end{aligned} \tag{11.44}$$

where y_α denote the 9 (spherical) coordinates and σ_α the corresponding normalized distribution. The parameter c is chosen such that the rms-radii of σ and ρ are close to each other. Starting from homogeneously distributed random numbers ξ_α in $[0, 1]$, the densities σ_α are used to calculate the coordinates y_α of the nucleon configurations

$$\xi_\alpha = \int_0^{y_\alpha} \mathrm{d}y \sigma_\alpha(y). \tag{11.45}$$

Given a configuration of nucleons characterized by the random numbers ξ_α^n [via (11.45)], a new configuration ξ_α^{n+1} is generated by

$$\xi_\alpha^{n+1} = \xi_\alpha^n + (\eta_\alpha - 0.5)\delta_\alpha, \tag{11.46}$$

where η_α is a random variable ($0 \le \eta_\alpha \le 1$) and δ_α is a constant (parameter "width"). The variables ξ_α are assumed to be periodic (mod 1). The new configuration will be accepted provided

$$\frac{\rho(\boldsymbol{x}_1^{n+1}, \boldsymbol{x}_2^{n+1}, \boldsymbol{x}_3^{n+1})\sigma(\boldsymbol{x}_1^n, \boldsymbol{x}_2^n, \boldsymbol{x}_3^n)}{\rho(\boldsymbol{x}_1^n, \boldsymbol{x}_2^n, \boldsymbol{x}_3^n)\sigma(\boldsymbol{x}_1^{n+1}, \boldsymbol{x}_2^{n+1}, \boldsymbol{x}_3^{n+1})} \ge \bar{\xi}; \qquad 0 \le \bar{\xi} \le 1, \tag{11.47}$$

where $\bar{\xi}$ is a random number, and the multiplicity $\omega^{n+1} = 1$ is assigned to this configuration. Otherwise, the multiplicity ω^n is increased by one and a new configuration according to the rule (11.46) is generated. Equation (11.47) displays the importance of the choice of the auxiliary density σ; if σ is close to the actual density ρ, only few of the newly generated configurations are rejected. This procedure leads asymptotically to a set of configurations distributed according to $\rho(\boldsymbol{x}_1, \boldsymbol{x}_2, \boldsymbol{x}_3)$. To improve convergence of the procedure, configurations are recorded only after an initialization period (parameter "nnot"); furthermore, to reduce correlations between different configurations, not all configurations generated according to the process (11.46,47) are recorded (parameter "njump"). Finally, the whole procedure is repeated for various initial configurations.

Ground-state expectation values of any local observable are given by appropriate sums over configurations:

$$\begin{aligned}\overline{O} &= \int \mathrm{d}^3x_1\mathrm{d}^3x_2\mathrm{d}^3x_3\rho(\boldsymbol{x}_1, \boldsymbol{x}_2, \boldsymbol{x}_3)O(\boldsymbol{x}_1, \boldsymbol{x}_2, \boldsymbol{x}_3) \\ &= \frac{\sum_n \omega_n O(\boldsymbol{x}_1^n, \boldsymbol{x}_2^n, \boldsymbol{x}_3^n)}{\sum_n \omega_n},\end{aligned} \tag{11.48}$$

and, in particular, we have for the projectile–nucleus amplitude

$$f(\boldsymbol{k}', \boldsymbol{k}) = \overline{f}_{\mathbf{k}',\mathbf{k}}. \tag{11.49}$$

The calculations are performed for the following ground-state density:

$$\rho(\boldsymbol{x}_1, \boldsymbol{x}_2, \boldsymbol{x}_3) = \rho_0 \exp\left\{-\frac{1}{b^2}\left[\left(\sum_{i=1}^3 \boldsymbol{x}_i^2\right) + \left(\sum_{i=1}^3 \boldsymbol{x}_i\right)^2\right]\right\}, \tag{11.50}$$

with $b = 1.41$ fm [11.18]. The parameter c in (11.44) is chosen as $c = 0.5b\sqrt{3}$. The form factor corresponding to this density can be calculated analytically:

$$\begin{aligned}F(q) &= \int \mathrm{d}^3x_1\mathrm{d}^3x_2\mathrm{d}^3x_3\rho(\boldsymbol{x}_1, \boldsymbol{x}_2, \boldsymbol{x}_3)\frac{1}{4}\left(\sum_i^3 \mathrm{e}^{\mathrm{i}\mathbf{q}\mathbf{x}_i} + \mathrm{e}^{-\mathrm{i}\mathbf{q}\sum_i^3 \mathbf{x}_i}\right) \\ &= \exp\left(-\frac{3}{16}b^2q^2\right)\end{aligned} \tag{11.51}$$

and will be used in testing the stochastic method. The statistical errors in the evaluation are estimated according to

$$[\Delta F(q)]^2 = \left\{ \frac{1}{N_s} \sum_{n=1}^{N_s} F_n^2(q) - \left[\frac{1}{N_s} \sum_{n=1}^{N_s} F_n(q) \right]^2 \right\} \frac{1}{N_s},$$

where N_s is the number of initial configurations (parameter "nstart") and $F_n(q)$ the form factor averaged over the configurations corresponding to the nth initial configuration. To display the fluctuations, the program evaluates the form factor for three orthogonal directions of the momentum transfer.

To illustrate the flexibility of the stochastic method with respect to the choice of ground-state density, the program offers the possibility of introducing a hard-core correlation that selects only those configurations with inter-nucleon distances greater than the hard-core radius r_c. The density (11.50) is accordingly modified:

$$\rho(\boldsymbol{x}_1, \ldots, \boldsymbol{x}_4) = \delta\left(\sum_{i=1}^{4} \boldsymbol{x}_i\right) \rho(\boldsymbol{x}_1, \boldsymbol{x}_2, \boldsymbol{x}_3) \prod_{i<j} \theta[(\boldsymbol{x}_i - \boldsymbol{x}_j)^2 - r_c^2]. \tag{11.52}$$

In this case, the Gaussian parameter b has to be readjusted such that the rms-radius remains unchanged.

Optical-Potential Calculation

This calculation is straightforward. The radial Schrödinger equation (11.39) is integrated with Milne's method [11.19]; the potential is [cf. (11.40,43,50)]

$$u(r) = \frac{-32}{\sqrt{3\pi}} \frac{a}{b^3} \exp\left(-\frac{4r^2}{3b^2}\right) . \tag{11.53}$$

Bessel and Hankel functions for the evaluation of the partial-wave scattering amplitudes f_l (11.41) and Legendre polynomials for evaluation of the total amplitude f are calculated using standard recursion relations [11.15].

11.2.3 Program Description MUSH (Multiple Scattering of Hadrons) and Performance

This program calculates multiple scattering on ^{4}He at arbitrary energy. It contains four independent parts that can be reached from a common menu. Input and output are described in the table.

Before performing extensive calculations, the user should estimate the calculation time with short runs. Care has to be taken with the use of the parameters *width* and *hard-core radius*, since inappropriate values can lead to very long calculational times.

The calculations were performed on an Olivetti M28 (80286 central processor and 80287 arithmetic coprocessor). In the multiple-scattering model, the generation of configurations using the parameters of Table 11.1 takes $\approx$ 3000 s. With this set of configurations the form factor is calculated in $\approx$ 1800 s, the scattering amplitude for momenta $k < 0.2$ fm^{-1} in $\approx$ 600 s, for values $k \geq 0.2$ fm^{-1} in $\approx$ 2500 s. The optical-potential calculation is carried out in $\approx$ 7 s.

Table 11.1. Sample parameters to be used in the program MUSH. The configurations are generated with Gaussian single-particle densities

	Variables		
	Name	Type	Value
Program selection in the opening menu	nans	integer	
Generation of configurations			
Number of initial configurations c_0	nstart	integer	2500
Number of configurations generated before accepting the first	nnot	integer	20
Number of configurations kept after each c_0	nkeep	integer	4
Number of configurations left out between two accepted configurations	njump	integer	5
Width δ in the variation (11.46)	width	real	0.2
Hard-core radius (fm) (11.52)	core	real	0.
Output on file CONFIG: *nstart*, *nkeep*, and the configurations			
Calculation of the form factor			
Number of values of momentum transfer q	moment	integer	21
q_{min} (fm^{-1})	lbound	real	0.
q_{max} (fm^{-1})	ubound	real	4.
Output on file FORM: $F(q)$; $\Delta F(q)$; $\sqrt{< r^2 >}$; number of all, (number of different/number of all) configurations			
Calculation of differential cross section			
Real part of the scattering length (fm)	s1	real	−1.
Imaginary part of the scattering length (fm)	s2	real	1.
Number of momentum values k of the incident projectile	knumb	integer	1
k_{min} (fm^{-1})	qmin	real	1.
k_{max} (fm^{-1}) (only if more than one momentum value)	qmax	real	–
angular step $\Delta\theta$ (degree)	nthet	integer	6
Output on file STOCHASTIC: scattering length, momenta; $\mathrm{d}\sigma/\mathrm{d}\Omega = \vert f(\boldsymbol{k}',\boldsymbol{k})\vert^2$, $f(\boldsymbol{k}',\boldsymbol{k})$, $\sqrt{< r^2 >}$; number of all, (number of different/number of all) configurations			
Optical-potential calculation			
Real part of the scattering length (fm)	a1	real	−1.
Imag. part of the scattering length (fm)	a2	real	1.
Momentum of the projectile (fm^{-1})	q	real	1.
Number of partial waves to be calculated	lmax	integer	7
Output on file OPTICAL: $f_l(k)$, $f(\boldsymbol{k}',\boldsymbol{k})$; $\mathrm{d}\sigma/\mathrm{d}\Omega = \vert f(\boldsymbol{k}',\boldsymbol{k})\vert^2$, q, lmax, scattering length			

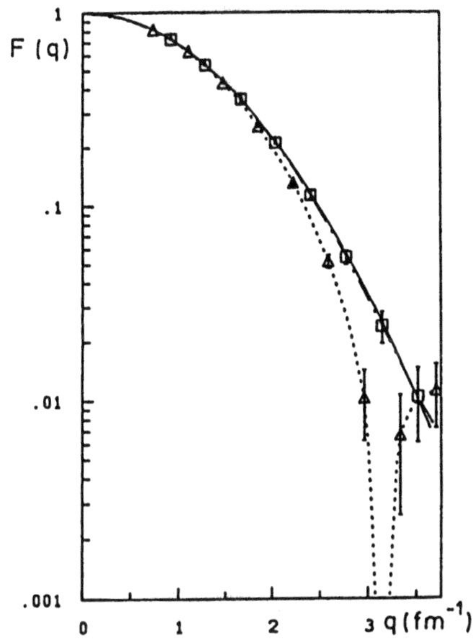

Fig. 11.1. Stochastic evaluation of the form factor of ^{4}He (*dot-dashed line*) in comparison with the exact result (11.53) (*solid line*); stochastic result for form factor corresponding to the density (11.52) with $b = 1.31$ fm; $r_c = 1.0$ fm (*dotted line*)

11.2.4 Discussion of the Results

Figures 11.1–2 display the results of the multiple-scattering calculations obtained with the program described above. The comparison in Fig. 11.1 between exact and stochastic evaluation of the form factor displays the accuracy obtainable with modest effort. Inclusion of a hard-core repulsion (see Fig. 11.1) – without change in the rms-radius – makes the form factor fall off more steeply, reflecting the suppression of the central density. Analytical evaluation of such a more realistic form factor [11.18] is not possible. The satisfactory agreement between exact and stochastic evaluation is relevant also for the stochastic evaluation of multiple-scattering processes, where destructive interference among different configurations in the process of configurational averaging is weaker than in the form factor. The calculated projectile–nucleus scattering length in Fig. 11.2(a) shows that for weak elementary amplitudes ($a \to 0$), both the optical-model approximation and the stochastic evaluation of the multiple-scattering equations approach the value $f(0) = 4a$. The strong-coupling limit of the stochastic and optical-model results is quite different. The optical-model calculation gives rise to hard-core scattering with an essentially real amplitude; real and imaginary parts of the full multiple-scattering amplitude evaluated stochastically are of the same or-

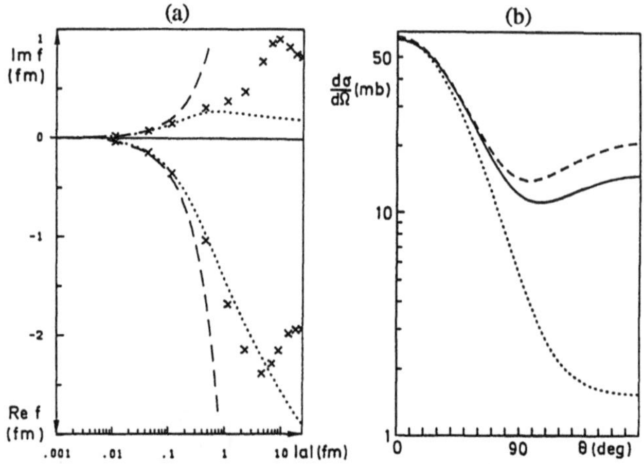

Fig. 11.2. (**a**) Real and imaginary part of the projectile–nucleus scattering length as a function of $|a|$, the absolute value of the projectile–nucleon scattering length (phase of a is 153°). *Dashed curve*: single-scattering approximation; *dotted curve*: optical-potential approximation; *crosses*: stochastic evaluation. (**b**) Differential cross section at $k = 1\ \mathrm{fm}^{-1}$, $a = (-1+\mathrm{i}\cdot 1)$ fm. *Solid line*: stochastic evaluation with hard-core radius $r_c = 0$; *dashed line*: the same with $r_c = 1.0$ fm; *dotted line*: optical-potential approximation

der of magnitude. This interesting strong-coupling limit might be interpreted as a trapping of the projectile by a certain class of configurations [11.8] and exhibits similarities with the localization limit in random media (localization also arises as a result of configurational averaging) [11.20]. Such trapping processes could be important in the strong absorption situation encountered in K^- and $\bar{p}$ atoms where the (absolute) values of the scattering lengths range from 1.5 to 4 fm [11.13]. At finite energies, this strong-coupling limit is reflected in the shape of the angular distribution. Trapping of the projectile leads – as might be expected – to an essentially isotropic distribution. This tendency is clearly seen in the comparison of the optical-model and stochastic evaluations in Fig. 11.2(b).

We have discussed here only a few of the many questions that can be investigated with this program. In particular, we advise the user to study convergence of the stochastic method by increasing the number of configurations for various incident-projectile momenta k and to investigate further the strong-coupling limit by performing the calculations for different choices of the phase of the elementary amplitude a. Furthermore, the method is not restricted to simple forms for the ground-state wave function. The functional form of the ground-state density is easily modified and thereby the effects of ground-state correlations can be investigated. Finally, other quantities of interest can be straightforwardly calculated, such as the average projectile wave function and Green's function or the summed elastic and inelastic cross sections by averaging $|f_{k',k}|^2$.

Our discussion has only described an exploratory study of a stochastic evaluation of multiple-scattering processes. Further investigations are necessary for developing this approach to allow finally for detailed comparison of results with experimental data (see [11.21] for such a comparison in the case of π–^{4}He scattering).

11.3 Hadronic Interactions in the Quark Model

11.3.1 Definition of the Model

One of the fundamental aims of nuclear physics, as we have said, is to understand hadronic interactions in terms of microscopic degrees of freedom, the quarks and gluons. Within the framework of non-relativistic quantum mechanics, the gluons are supposed to be eliminated, and the interaction of quarks and antiquarks is described by potentials. The defining property of all quark model descriptions of hadronic systems is the confinement of the elementary constituents. This property is straightforwardly implemented in the description of single hadrons. Serious problems are encountered in the attempt to formulate a non-relativistic many-body theory with confining forces for the description of interacting hadrons. Within conventional dynamics, such attempts cannot be successful. Rather, the principle of superposition of pair energies yielding the total potential energy has to be abandoned. To avoid confinement of hadrons or long-range residual interactions, saturation of the confining interaction by many-body forces is required. These properties of the interaction are realized in the following ansatz for the potential energy of N quarks (coordinates $\boldsymbol{r}_i$) and N antiquarks ($\boldsymbol{r}_{\bar{j}}$):

$$V = \min_s \sum_{i=1}^{N} U[\boldsymbol{r}_i - \boldsymbol{r}_{\bar{s}(i)}], \tag{11.54}$$

where $U(\boldsymbol{r})$ is the confining potential of a single-quark–antiquark pair and the permutation $\bar{s}(i)$ specifies the linking of quarks to antiquarks. Saturation of the confining interaction is obvious; each quark interacts only with one antiquark. The partner with which this interaction takes place is for a given configuration chosen such that the total potential energy is minimal. Interaction between different pairs occurs, since in the time evolution of the system, the quark and antiquark exchange their partner, that is, the basic and, at the level of this model calculation, only mechanism to generate hadronic interactions is quark exchange. Matching of quark and antiquark coordinates depends on the multiquark configuration. The potential energy is not the superposition of the $N(N-1)/2$ pair energies. Thus, many-body forces as a reflection of the underlying non-linear quark-gluon dynamics constitute an essential element in the definition of this model.

The simplest system relevant for a study of hadronic interactions on the basis of quark degrees of freedom is the four-body system, containing two quarks and two antiquarks, describing two interacting mesons. For this system, the following properties are direct consequences of the definition (11.54) of the interaction. The potential energy becomes infinite when the coordinates of a single quark or antiquark go to infinity:

$$V \overset{r_i \to \infty}{\longrightarrow} \infty . \tag{11.55}$$

Secondly, the potential energy reduces to that of quarks and antiquarks for non-interacting mesons, if simultaneously quark and antiquark coordinates approach infinity:

$$V \overset{r_{12} \to \infty}{\longrightarrow} \begin{cases} U(\boldsymbol{r}_{1\bar{1}}) + U(\boldsymbol{r}_{2\bar{2}}) & r_{1\bar{1}} \ \text{finite} \\ U(\boldsymbol{r}_{1\bar{2}}) + U(\boldsymbol{r}_{2\bar{1}}) & r_{1\bar{2}} \ \text{finite}, \end{cases} \tag{11.56}$$

with ($\boldsymbol{r}_{i\bar{j}} = \boldsymbol{r}_i - \boldsymbol{r}_{\bar{j}}$); that is, quarks are confined and hadrons at large distances are non-interacting. In general, four-body problems also in the presence of confining interactions cannot be solved exactly, and the resulting dynamics between hadrons has to be deduced within more or less well-defined approximation schemes. With harmonic confinement, the model (11.58) can be sufficiently simplified by analytical means to allow for an exact numerical solution. For the following, we assume that

$$U(\boldsymbol{r} - \boldsymbol{r}') = \frac{1}{4} m\omega^2 (\boldsymbol{r} - \boldsymbol{r}')^2 . \tag{11.57}$$

Thus, V of (11.54) is written explicitly as

$$\begin{aligned} V &= \frac{1}{4} m\omega^2 \left[\left(\boldsymbol{r}_{1\bar{2}}^2 + \boldsymbol{r}_{2\bar{1}}^2 \right) \theta \left(\boldsymbol{r}_{1\bar{1}}^2 - \boldsymbol{r}_{1\bar{2}}^2 + \boldsymbol{r}_{2\bar{2}}^2 - \boldsymbol{r}_{2\bar{1}}^2 \right) \right. \\ &+ \left. \left(\boldsymbol{r}_{1\bar{1}}^2 + \boldsymbol{r}_{2\bar{2}}^2 \right) \theta \left(\boldsymbol{r}_{1\bar{2}}^2 - \boldsymbol{r}_{1\bar{1}}^2 + \boldsymbol{r}_{2\bar{1}}^2 - \boldsymbol{r}_{2\bar{2}}^2 \right) \right] , \end{aligned}$$

and the Hamiltonian of the $2q - 2\overline{q}$ system is

$$H = \frac{1}{2m} \left(p_1^2 + p_2^2 + p_{\bar{1}}^2 + p_{\bar{2}}^2 \right) + V . \tag{11.58}$$

11.3.2 Solution of the 4-Quark Problem

The following method rests crucially upon the property of confinement that allows asymptotically for two-body channels only, that is, the asymptotic configurations consist of two mesons in their respective ground or excited states. This allows formulation of the scattering problem entirely in terms of square-integrable functions and solution of the problem with harmonic confinement by reduction to a one-dimensional integral equation.

As a first step towards the solution, the following set of coordinates is introduced:

$$\boldsymbol{R} = \frac{1}{2}(\boldsymbol{r}_1 + \boldsymbol{r}_2 + \boldsymbol{r}_{\bar{1}} + \boldsymbol{r}_{\bar{2}}), \qquad \boldsymbol{x} = \frac{1}{2}(\boldsymbol{r}_1 + \boldsymbol{r}_2 - \boldsymbol{r}_{\bar{1}} - \boldsymbol{r}_{\bar{2}}),$$
$$\boldsymbol{y} = \frac{1}{2}(\boldsymbol{r}_1 + \boldsymbol{r}_{\bar{1}} - \boldsymbol{r}_2 - \boldsymbol{r}_{\bar{2}}), \qquad \boldsymbol{z} = \frac{1}{2}(\boldsymbol{r}_1 - \boldsymbol{r}_{\bar{1}} - \boldsymbol{r}_2 + \boldsymbol{r}_{\bar{2}}). \tag{11.59}$$

This orthogonal transformation yields

$$\begin{aligned} H &= -\frac{\hbar^2}{2m}\left(\frac{1}{4}\Delta_R + \Delta_x + \Delta_y + \Delta_z\right) \\ &+ \frac{1}{2}m\omega^2\left[x^2 + y^2\theta(z^2 - y^2) + z^2\theta(y^2 - z^2)\right]. \end{aligned} \tag{11.60}$$

Equation (11.60) displays decoupling not only of the motion of the center of mass $\boldsymbol{R}$ of the four-body system but also of the confined "dipole" variable $\boldsymbol{x}$. The channel variables $\boldsymbol{y}$ and $\boldsymbol{z}$ are coupled in the potential energy. The motion in the larger of the two variables is free, while the smaller one is subject to confinement. The transition from $y > z$ to $y < z$ corresponds, as the definition (11.59) shows, to rearrangement of the quarks from the $(1,\bar{1})$, $(2,\bar{2})$ into the $(1,\bar{2})$, $(2,\bar{1})$, configuration. The coupling between the $\boldsymbol{y}$ and $\boldsymbol{z}$ degrees of freedom is independent of their relative orientation; as a result, the associated angular momenta are separately conserved and the "radial" Schrödinger equation for the y–z motion, describing the scattering of two s-wave mesons with corresponding quark–antiquark s-wave functions, reads

$$\begin{aligned} \Bigg\{\left(\frac{-\mathrm{d}^2}{\mathrm{d}z^2} + \frac{-\mathrm{d}^2}{\mathrm{d}y^2}\right) &+ \frac{m^2\omega^2}{\hbar^2}\left[z^2\theta(y^2 - z^2) + y^2\theta(z^2 - y^2)\right] \\ &- k_0^2 + \frac{2m}{\hbar^2}\epsilon_0\Bigg\}\psi_0(y,z) = 0, \end{aligned} \tag{11.61}$$

$$\epsilon_n = \hbar\omega(2n + 3/2) = E - \frac{\hbar^2 k_n^2}{2m}. \tag{11.62}$$

ϵ_n represents the energy corresponding to the internal motion of the quarks. We observe that the Schrödinger equation (11.61) is symmetric under exchange of the y and z variables, reflecting the definite exchange symmetry of the original Hamiltonian under quark and antiquark exchange, respectively. Therefore, the problem can be reduced further by restricting the values of y and z to the triangle Δ defined by $0 \le z \le y$ and requiring appropriate boundary conditions at the rearrangement surface $y = z$. The wave function satisfying (11.61) is specified completely by the following boundary conditions:

1. The standard regularity condition for $z \to 0$

$$z \to 0: \qquad \psi_m^{s,a}(y,z) \longrightarrow z. \tag{11.63}$$

2. The incoming wave condition for $y \to \infty$

$$y \to \infty: \quad \psi_m^{s,a}(y,z) \propto \mathrm{e}^{-\mathrm{i}k_m y}\Phi_m(z) - \sum_n S_{n,m}^{s,a}\mathrm{e}^{\mathrm{i}k_n y}\Phi_n(z). \tag{11.64}$$

3. The requirement of symmetry or antisymmetry

$$\text{symmetric case:} \qquad (\partial_y - \partial_z)\psi_m^s(y,z)\big|_{y=z} = 0, \tag{11.65}$$
$$\text{antisymmetric case:} \qquad \psi_m^a(y,z)\big|_{y=z} = 0. \tag{11.66}$$

The boundary condition (11.64) specifies the incoming two-meson state as a product state of a harmonic-oscillator wave function for the internal motion of the quarks and the wave function for the free relative motion of the two mesons. The functions

$$\tilde{\Phi}_n^{\pm}(y,z) = \mathrm{e}^{\pm\mathrm{i}k_n y}\Phi_n(z) \tag{11.67}$$

satisfy the equation

$$\begin{aligned} h(y,z)\tilde{\Phi}^{\pm}(y,z) &= \left[-\frac{\hbar^2}{2m}\left(\frac{\partial^2}{\partial y^2} + \frac{\partial^2}{\partial z^2}\right) + \frac{1}{2}m\omega^2 z^2\right]\tilde{\Phi}^{\pm}(y,z) \\ &= E\tilde{\Phi}^{\pm}(y,z)\,. \end{aligned} \tag{11.68}$$

The asymptotic wave function (11.64) contains a sum over outgoing waves corresponding to the various excited two-meson states. The weights of these components in the asymptotic wave function are the S-matrix elements $S_{n,m}^{s,a}$.

11.3.3 Calculational Methods

In the course of these simplifications, we have converted the four-body Schrödinger equation corresponding to the Hamiltonian (11.58) to a two-dimensional boundary-value problem, which, by applying techniques well known from electrostatics, is further reduced to a system of algebraic equations. The key observation is that both the still-unknown wave functions $\psi_m^{s,a}(y,z)$ and the function $\tilde{\Phi}^{\pm}(y,z)$ defined in (11.68) satisfy in the triangle Δ the same wave equation (11.68). These functions differ in their behavior at the boundaries. As a consequence, Green's identity [11.22] can be used to derive

$$\begin{aligned} 0 &= \int_\Delta \mathrm{d}\tau \left[\tilde{\Phi}_l^{\pm} h(y,z)\psi_m^{s,a} - \psi_m^{s,a} h(y,z)\tilde{\Phi}_l^{\pm}\right] \\ &= \int_{\delta\Delta} \mathrm{d}\sigma \left[\tilde{\Phi}_l^{\pm}(-\partial_n)\psi_m^{s,a} - \psi_m^{s,a}(-\partial_n)\tilde{\Phi}_l^{\pm}\right], \end{aligned} \tag{11.69}$$

where $\delta\Delta$ is the border of the triangle consisting of the pieces $y = z; z = 0$ and the asymptotic boundary $y = R; R \to \infty$.

Evaluating the surface integral in (11.69) yields equations to determine the wave functions or their normal derivatives along the boundary $y = z$:

$$-2\mathrm{i}\delta_{lm} = \int_0^\infty \mathrm{d}t\, \left[\tilde{\Phi}_l^+(-\partial_z + \partial_y)\psi_m^{s,a} - \psi_m^{s,a}(-\partial_z + \partial_y)\tilde{\Phi}_l^+\right]\Big|_{y=z=t}, \quad (11.70)$$

which, using (11.65) and (11.66), can be written as

$$\delta_{lm} = \int_0^\infty \mathrm{d}t\ \chi_l^+(t)\omega^m(t); \qquad \delta_{lm} = \int_0^\infty \mathrm{d}t\ \Phi_l^+(t)\theta^m(t), \quad (11.71)$$

where the following functions on the diagonal $y = z$ have been introduced:

$$\Phi_l^\pm(t) = \tilde{\Phi}_l^\pm(t,t), \qquad \chi_l^\pm = (\partial_z - \partial_y)\tilde{\Phi}_l^\pm(y,z)\Big|_{y=z=t},$$
$$\omega^m(t) = -\frac{\psi_m^s(t,t)}{2\mathrm{i}k_m}, \qquad \theta^m(t) = \frac{(\partial_z - \partial_y)\psi_m^a(y,z)}{2\mathrm{i}k_m}\Bigg|_{y=z=t}. \quad (11.72)$$

The system of algebraic equations (11.71) determines the normal derivatives and wave functions on the diagonal for symmetric and antisymmetric scattering, respectively. Knowledge of the wave function and normal derivative on the diagonal is sufficient to determine the wave functions in the whole triangle by continuing as in electrostatics, via Green's function techniques, the wave function from the boundary into the interior of the triangle. The S-matrix elements can be directly expressed in terms of the diagonal functions defined in (11.72), as can be immediately deduced from (11.69). The result is

$$S_{lm}^s = \frac{k_m}{k_l}\int \mathrm{d}t\chi_l^-(t)\omega^m(t), \qquad S_{lm}^a = \frac{k_m}{k_l}\int \mathrm{d}t\Phi_l^-(t)\theta^m(t). \quad (11.73)$$

Thus, the original Schrödinger equation (11.61) has been reduced to the algebraic problem of constructing sets of functions ω_m, θ_m biorthogonal to the given sets of functions χ_l^+ and $\tilde{\Phi}_l^+$, respectively. The biorthogonalization procedure is performed by expansion of the functions ω^m and θ^m in terms of appropriately chosen sets of functions. A particularly convenient and numerically efficient choice is the following:

$$\omega^m(t) = \sum_i a_{mi}\Phi_i^+(t), \quad \theta^m(t) = \sum_i b_{mi}\chi_i^+(t). \quad (11.74)$$

For this choice, (11.71) implies $b_{mi} = a_{mi}$, that is, the two sets of functions ω^m, θ^m are obtained by one matrix inversion:

$$\sum_i a_{im}\int_0^\infty \mathrm{d}t\ \chi_l^+(t)\Phi_i^+(t) = \delta_{lm}. \quad (11.75)$$

The calculations are further simplified by applying Green's identity as used in (11.70) to pairs of functions of the set $\tilde{\Phi}_i^\pm(y,z)$. The following identities are straightforwardly derived:

$$\int_0^\infty \mathrm{d}t\ \chi_l^\pm(t)\Phi_i^\pm(t) = \int_0^\infty \mathrm{d}t\ \Phi_l^\pm(t)\chi_i^\pm(t), \quad (11.76)$$
$$\int_0^\infty \mathrm{d}t\ \chi_l^-(t)\Phi_i^+(t) = \int_0^\infty \mathrm{d}t\ \Phi_l^-(t)\chi_i^+(t) + 2\mathrm{i}k_i\delta_{il}. \quad (11.77)$$

As a consequence, symmetric and antisymmetric S-matrix elements are related [cf. (11.73)]

$$S^a_{lm} = S^s_{lm} - 2\mathrm{i}k_m a_{lm}. \tag{11.78}$$

Finally, calculation of the relevant matrix elements in (11.71) is simplified using

$$\int_0^\infty \mathrm{d}t\ \chi_l^+(t)\Phi_i^+(t) = -\mathrm{i}(k_l + k_i)\int_0^\infty \mathrm{d}t\ \Phi_l^+(t)\Phi_i^+(t), \tag{11.79}$$

as follows from the definitions (11.72) and the symmetry relation (11.76). Summarizing, the expansion coefficients a_{in} are obtained from

$$\delta_{ln} = \sum_m A_{lm} a_{mn}\,, \tag{11.80}$$

with

$$A_{lm} = -\mathrm{i}(k_l + k_m)\int_0^\infty \mathrm{d}t\ \mathrm{e}^{\mathrm{i}(k_l+k_m)t}\Phi_l(t)\Phi_m(t). \tag{11.81}$$

By continuation of the channel momenta k_l into the complex plane, these expressions can be used to determine the singularities of the S-matrix that arise for det $A = 0$.

11.3.4 Numerical Methods

For the calculations, the following set of constants is chosen: $\hbar = 2m = 1$; $\omega = 4$. In these units, the extension of a meson is given by the rms-radius $\bar{R} = \sqrt{3/8(2\hbar)/(m\omega)} = \sqrt{3/8}$, and the relation between the channel momenta corresponding to the various states that can be excited in s-wave scattering reads $k_n^2 = k_0^2 - 8n$ (that is, the first inelastic threshold occurs at $k_0 = \sqrt{8}$). The harmonic-oscillator eigenfunctions are expressed by the Laguerre polynomials,

$$\Phi_l(z) = N_l z \mathrm{e}^{-z^2} L_l^{1/2}(2z^2), \quad N_l^2 = 4\sqrt{2}\frac{\Gamma(l+1)}{\Gamma(l+\frac{3}{2})}, \tag{11.82}$$

and calculated using standard recursion relations for the $L_l^{1/2}$ [11.19]. The integrations for the calculation of the matrix elements A_{lm} (11.81) are performed with the Romberg algorithm [11.16] with non-equidistant mesh points. The results for a_{mn} (11.80) are rather sensitive to both inaccuracies in the numerical integration and inaccuracies of the inversion procedure. This sensitivity arises since, with increasing number of basis states N and therefore increasing dimension of A, det A approaches zero very fast. We calculate approximatively det A by replacing Φ_l in (11.81) by the large-l asymptotic values [11.19]

$$\Phi_l(t) = \frac{4}{N_l}\sqrt{\frac{2}{\pi(2l+3/2)}}\cos\left(2t\sqrt{2l+3/2} - \frac{\pi}{2}\right) \tag{11.83}$$

and obtain

$$A_{lm} = \frac{8}{\pi N_l N_m} \frac{1}{l + m + 3/4 - k_0^2}. \tag{11.84}$$

Using the Gaussian integral representation for the determinant of the $N \times N$-matrix $\bar{A}(N)$ with matrix elements

$$\bar{A}_{lm} = \frac{1}{\alpha_l + \alpha_m}, \tag{11.85}$$

a recursion relation with respect to the dimension N is easily derived:

$$\det\left[\bar{A}(N)\right] = \det\left[\bar{A}(N-1)\right] \prod_{i=1}^{N-1} \left(\frac{\alpha_N - \alpha_i}{\alpha_N + \alpha_i}\right)^2. \tag{11.86}$$

Collecting the results, we have

$$\det\left[A(N)\right] = \left(\frac{4}{\pi}\right)^N \prod_{l=1}^{N} \frac{1}{N_l^2(l + \frac{3}{8} - k_0^2/2)} \prod_{i \le j} \frac{(i-j)^2}{(i + j + \frac{3}{4} - k_0^2)^2}. \tag{11.87}$$

It is easy to see that det A decreases with the number N of basis states faster than $N!$. Therefore, an increase in the number of basis states without precaution does not necessarily yield higher accuracy. Since this possible loss in accuracy arises from the large-l states, the energetically closed channels with real matrix elements, we decompose A into two real matrices C, D with $A = C + \mathrm{i}D$ and invert C with double precision. The inversion involving the few open channels giving rise to complex matrix elements is performed without problem:

$$A^{-1} = \left(1 + \mathrm{i}C^{-1}D\right)^{-1} C^{-1}.$$

In searching for the pole position of the S-matrix in the complex energy plane, no such decomposition of A is possible, since in general the channel momenta are complex. We therefore recommend use of a small number of basis states, unless the evaluation of the complex determinant is performed with double precision.

11.3.5 Program Description QUASH (Quarks and Hadron Scattering) and Performance

The program calculates the diagonal S-matrix elements in the "0-channel" and, if energetically possible, the "1-channel" [see (11.73)] both for symmetric and antisymmetric scattering.

With the values of the parameters of the calculation given in Table 11.2, the computational time is $\approx$ 30 s for det A and $\approx$ 40 s for the S-matrix elements at a single momentum value.

Table 11.2. Sample parameters for the program QUASH

	Variables		
	Name	Type	Value
Choice between built-in and redefined parameters for integration	nans	integer	
If parameters are redefined:			
Accuracy	eps	real	1E−10
Upper integration limit	ulimit	real	8.
Number of subintervals	interv	integer	20
Calculation of det A (11.81)			
Number of basis states	lmax	integer	8
Real part of energy	e1	real	−0.5–3.
Imaginary part of energy	e2	real	0.–0.5
Output: value of det A on screen			
Other values of energy	nans	integer	
Calculation of S-matrix elements			
Number of basis states	lmax	integer	8
Number of values of the momentum k	nk0	integer	≥ 0
Lower limit of k	k0min	real	$\geq 0.$
Upper limit of k	k0max	real	$\leq 6.$
Output on files SYM00,SYM11,ASYM00,ASYM11: parameters of the calculation and the corresponding S-matrix elements			

11.3.6 Discussion of the Results

Figure 11.3 shows as a function of incident momentum the diagonal S-matrix elements, which as usual are parametrized in terms of phase shifts and inelasticity parameter

$$S_{ll} = \eta_{ll} e^{2i\delta_{ll}}.$$

A glance at the results makes the physical interest of the model obvious. Rather intricate and complex behavior of the S-matrix, reflecting the composite nature of the hadrons, is obtained despite the extreme simplicity of the calculation, which hardly exceeds that of a simple-potential-scattering calculation. Of particular interest is the emergence of the narrow resonances occurring at energies near the inelastic thresholds and the existence of a bound state with small binding energy $\epsilon = -0.16$. These results point to the appearance of a new energy scale, which may describe the change from typical particle-physics energies to nuclear-physics energies, and which is reminiscent of similar phenomena occurring in the e–H system where weak bound states and narrow resonances are also observed [11.23], indicating there the change from atomic to molecular energy scales. It is remarkable that in the context of a model of hadronic interactions narrow resonances occur, although no angular momentum barrier prevents the strong interaction from decomposing the resonating system ("super-allowed" decay). Narrow resonances [such as the $\rho(975)$ or the $a_0(980)$] close to the $K\bar{K}$ threshold are actually observed, and they may have their origin in quark exchange as described here. Less dramatic is the behavior of the repulsive antisymmetric scattering (Fig. 11.4). The repulsion arises as a result of the Pauli principle. Nevertheless, this rather smooth momentum dependence of δ_{00} is not of the kind that can be described by a potential-scattering model. For instance, the decrease of the phase shift by π contradicts the Levinson theorem for potential scattering [11.24] and reflects the composite nature of the hadrons. Properties of other S-matrix elements such as S_{11} that can also be obtained with this program have not been investigated systematically and are poorly understood. To become familiar with the numerical subtleties of this method, we first recommend studies of the convergence and accuracy of the method. With appropriate care, the calculation of the S-matrix elements can be extented to higher energies, and by trivial modifications of the program, calculations of S-matrix elements other than 0-0 and 1-1 are possible. We further recommend a study of the resonances by locating the poles in the complex plane and investigating quantitatively and systematically their widths and energies. This can be done using the part of the program in which det A is calculated. Further insight into the dynamical origin of repulsion or attraction, bound state, and resonances can be obtained by replacing the harmonic-oscillator dependence z^2 in (11.61) by other confining potentials, such as a linear potential $|z|$ or an infinite square-well potential. This is achieved by replacing in the definition of the matrix A the harmonic-oscillator wave functions by the corresponding bound-state wave functions (with corresponding changes in the eigenenergies). Finally, within the same computational scheme, internal color degrees of freedom can be included as described in [11.12].

11.4 Conclusion

We have described two problems that illustrate important directions of research in the physics of strong interactions. We have chosen to discuss, on the

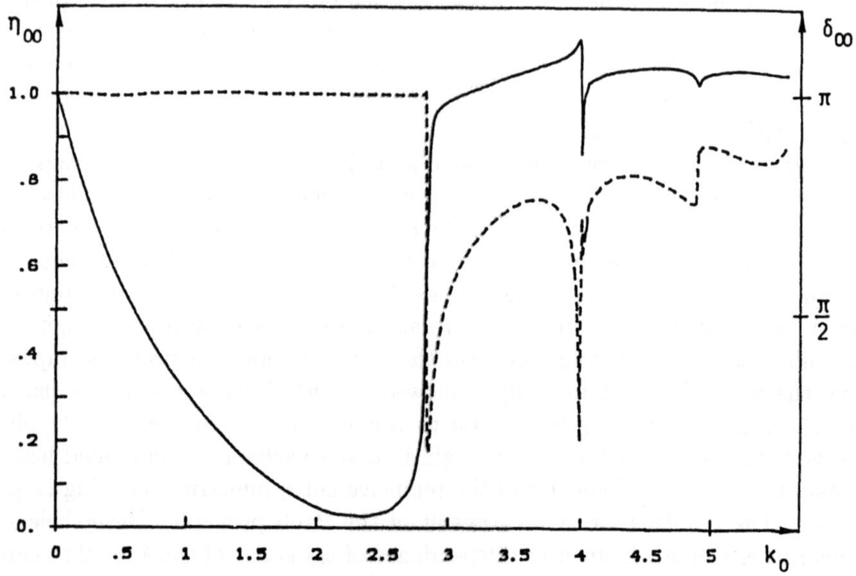

Fig. 11.3. Inelasticity parameter (*dashed line*) and phase shift (*solid line*) for symmetric scattering in the 0–0 channel

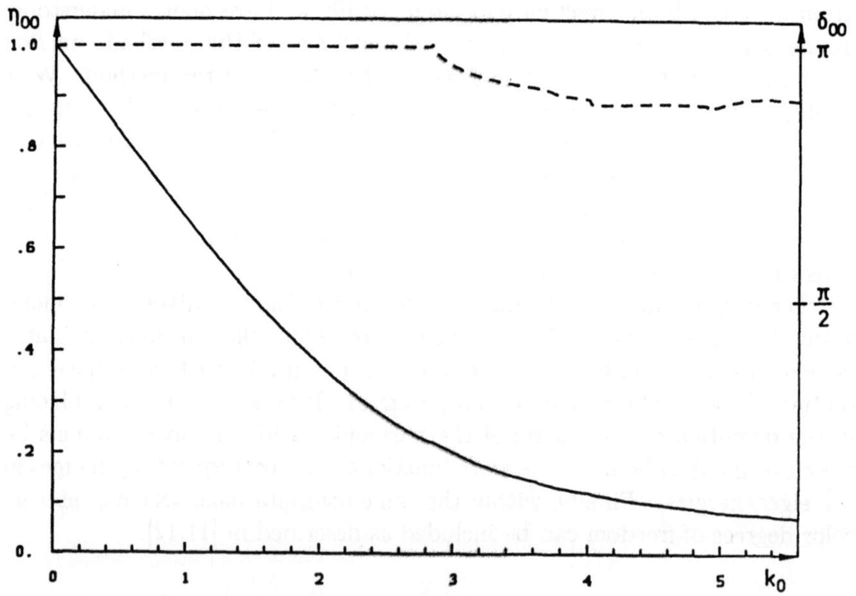

Fig. 11.4. Inelasticity parameter (*dashed line*) and phase shift (*solid line*) for antisymmetric scattering in the 0–0 channel

one hand, the multiple-scattering formalism, which provides the basic framework for practically all numerical studies of high-energy nuclear reactions. We have addressed, on the other hand, the issue of quark-model formulations of hadronic interactions, which are fundamental to nuclear physics. In both cases, our discussion has focused on basic concepts and methods. New developments have been particularly emphasized.

The multiple-scattering approach to nuclear reactions has a wide range of applications. We have discussed specifically the optical-potential approximation, which forms the basis of most of the traditional numerical applications. The optical-potential approximation is the first term of a systematic low-density expansion. Until today, and despite many efforts, technical limitations have prevented complete calculation of the second-order correction. Related technical difficulties appear in corresponding descriptions of inelastic processes such as two-nucleon knock-out or inclusive reactions. To match the increasing possibilities in experimental techniques, in particular as far as investigations of multiparticle reactions are concerned, development of new numerical approaches in theoretical studies is necessary. The stochastic approach constitutes such an attempt towards a more complete treatment of multiple-scattering processes. Our discussion has demonstrated the feasibility of a stochastic evaluation of multiple-scattering processes in the context of a simple model problem, and it is hoped that it will initiate further developments. Study of such new computational methods will inevitably also improve the analytical understanding of the multiple-scattering formalism. Already at the level of the simple model calculation presented here, the stochastic method in comparison with the optical-potential approximation has revealed the existence of an unexpected strong-coupling limit of the multiple-scattering equations. This limit is theoretically interesting; it is a phenomenon similar to localization in disordered media. It could be of practical relevance for the explanation of the poorly understood low-energy strong-absorption situation as encountered, for instance, in K^- or $\bar{p}$ atoms. However, for such specific and detailed applications of the stochastic methods to be reliable, further developments are required; most importantly, the fixed-scatterer approximation underlying the model calculations presented has to be relaxed.

Our discussion of the quark-model description of hadronic interactions has focused on a very different aspect of computational nuclear physics. Here, the problems to be investigated are of a qualitative nature, such as the origin of the short-range repulsion and medium-range attraction in the NN system or the possible existence of resonances in meson–meson scattering. Numerical methods are employed for a study of these qualitative problems; these models do not provide a basis for detailed quantitative investigations. The need for numerical approaches arises since in the quark model the simplest hadronic system is a four-body problem. In studies of the specific problems encountered in formulating a many-body theory with confined constituents, a new class of numerically exactly solvable four-body problems has been found. We have described here the new methods of solution that these models offer, owing to the presence of confining interactions. These methods allow exact evaluation of the corresponding non-trivial many-channel S-matrix. Apart from the general interest in exact solutions of a four-body problem and the formal interest in the application of new methods, the specific physical interest concerns the relation between the simple underlying dynamics of the constituents and the resulting complex hadronic phenomena. The results of these model studies are reminiscent of many of the observed properties of the strong interaction and encourage further investigations and applications of these models and the methods of solutions. In this context, extension of the numerical studies to larger systems is particularly interesting. Exploratory investigations have been performed for the six-quark system [11.25] and for hadronic matter in one dimension [11.26]. The unusual form (11.54) of the potential energy provides the basis for such investigations within a well-defined theoretical framework, but it requires the development of new analytical and numerical methods.

We have exemplified computational methods in the description of hadron–hadron and hadron–nucleus interactions. Owing to its introductory nature, our discussion did not address the comparison of the results of the methods with experimental data. Such comparisons are clearly the motivation for most of the computational efforts. Nevertheless, as our contribution has demonstrated, numerical investigations are equally important for the development of new approaches in both quantitative and qualitative studies in the physics of strong interactions.

11.5 Technical Note

The programs `MUSH` and `QUASH` run on a microcomputer. They expect input to be read from the keyboard. Sample values for the input parameters are given in Tables 11.1 and 11.2. Example output files `FORM`, `STOCHAST`, and `OPTICAL` for the code `MUSH` (evaluated for 1000 configurations) and `SYM00`, `ASYM00` for the code `QUASH` (using 10 basis states) are included on the disk.

We acknowledge Dr. T. Karapiperis' careful and critical reading of the manuscript.

References

[11.1] L.S. Rodberg and R.M. Thaler, *Introduction to The Quantum Theory of Scattering* (Academic, New York, 1967) Chapt. 12;
M.L. Goldberger and K.M. Watson, *Collision Theory* (Wiley, New York, 1964) Chapt. 11

[11.2] J.M. Eisenberg and D.S. Koltun, *Theory of Meson Interactions with Nuclei* (Wiley, New York, 1980) Chapts. 4,5

[11.3] C.J. Batty, *Proceedings of the International School of Physics with Low Energy Antiprotons* (Erice, 1988); *Proceedings of the Workshop on Hadron Facilities and Antiproton Physics* (Turino, 1989) [$\bar{p}$-atoms]
C.J. Batty, *Proceedings of the 12th International Conference on Few Body Problems in Physics* (Vancouver, 1989) [$\bar{p}, K^-$-atoms]

[11.4] M. Hirata et al., Ann. Phys. N.Y. **108**, 116 (1977); Ann. Phys. N.Y. **120**, 205 (1979);
F. Lenz, Prog. Theor. Phys. Suppl. **91**, 27 (1987)

[11.5] S.J. Wallace, *Advances in Nuclear Physics* Vol. 12, edited by J.W. Negele and E. Vogt (Plenum, New York, 1981) p.135

[11.6] K. Zalewski, Ann. Rev. Nucl. Part. Sci. (1985) p.55

[11.7] R.J. Glauber, *Lectures in Theoretical Physics* Vol. 1, edited by W.E. Brittin et al. (Interscience, New York, 1959) p.315
R.J. Glauber and G. Matthiae, Nucl. Phys B**21**, 135 (1970)

[11.8] D.Stoll, Diplomarbeit (Erlangen, September 1988, unpublished)

[11.9] A.W. Hendry and D.B. Lichtenberg, Fortschr. der Phys. **33**, 139 (1985)

[11.10] K. Shimizu, Rep. Prog. Phys. **52**, 1 (1989)

[11.11] K. Yazaki, Prog. Theor. Phys. Suppl. **91**, 146 (1987)

[11.12] F. Lenz et al. Ann. Phys. N.Y. **170**, 65 (1986)

[11.13] T. Takaki and M. Thies, Phys. Rev. C**38**, 2230 (1988)

[11.14] T. Karapiperis and M. Kobayashi, Ann. Phys. N.Y. **177**, 1 (1987)

[11.15] A. Messiah, *Quantum Mechanics* (North-Holland, Amsterdam, 1970)

[11.16] G. Engeln-Müllges and F. Reutter, *Formelsammlung zur numerischen Mathematik* (Bibliographisches Institut, Mannheim, 1988) 6th Edition

[11.17] M.H. Kalos and P.A. Whitlock, *Monte Carlo Methods* Vol. I *Basics* (Wiley, New York, 1986)

[11.18] C.W. de Jaeger et al., Atomic Data and Nuclear Data Tables **14**, 479 (1974)

[11.19] M. Abramowitz and I.A. Stegun (eds.), *Handbook of Mathematical Functions* (National Bureau of Standards, Washington DC, 1964)

[11.20] T.N. Ramakrishnan, *Chance and Matter*, edited by J. Soulitie et al. (North-Holland, Amsterdam, 1987); B. Souillard, ibid.

[11.21] W.R. Gibbs et al., Phys. Rev. C**13**, 2433 (1976)

[11.22] J.D. Jackson, *Classical Electrodynamics* (Wiley, New York, 1975) Chapts. 2, 3

[11.23] Y.K. Ho, Phys. Rep. **99**, 1 (1983)
G.J. Schulz, Rev. Mod. Phys. **45**, 378 (1973)

[11.24] R.G. Newton, *Scattering Theory of Waves and Particles* (Springer, New York, 1982) Chapt. 12

[11.25] T. Sato, Europhys. Lett. 4, 991 (1987)

[11.26] C.J. Horowitz et al., Phys. Rev. D**31**, 1689 (1985)

Subject Index

MIX
Papier aus verantwortungsvollen Quellen
Paper from responsible sources
FSC® C105338

If you have any concerns about our products,
you can contact us on
ProductSafety@springernature.com

In case Publisher is established outside the EU,
the EU authorized representative is:
Springer Nature Customer Service Center GmbH
Europaplatz 3, 69115 Heidelberg, Germany

Printed by Libri Plureos GmbH
in Hamburg, Germany